T0222734

The

# Geology of the Lake District

and the Scenery as influenced by
Geological Structure

Jonathan Otley

The

# Geology of the Lake District

and the Scenery as influenced by
Geological Structure

by

J. E. MARR, Sc.D., F.R.S.

FELLOW OF ST JOHN'S COLLEGE, CAMBRIDGE

Cambridge:
at the University Press
1916

# CAMBRIDGE
## UNIVERSITY PRESS

University Printing House, Cambridge CB2 8BS, United Kingdom

Cambridge University Press is part of the University of Cambridge.

It furthers the University's mission by disseminating knowledge in the pursuit of education, learning and research at the highest international levels of excellence.

www.cambridge.org
Information on this title: www.cambridge.org/9781107492820

© Cambridge University Press 1916

First published 1916
First paperback edition 2015

A catalogue record for this publication is available from the British Library

ISBN 978-1-107-49282-0 Paperback

# PREFACE

THE English Lake District, owing to the variety of its geological structure and its physical features, is well adapted to the requirements of the geological student. It is one of Nature's laboratories, where practical work may be successfully accomplished. An increasing number of those who are interested in geology visit it annually, and for the use of these visitors a work, in which the results of the various labours of those who have published writings upon the geology of the district are summarised, is needed.

I have endeavoured to write a condensed account of the geology of the area, and of the influence of geological structure upon its physical features, and I have added references to books and papers where detailed accounts may be obtained by those who wish to conduct further enquiries along special lines.

The work is primarily written for the student. By printing in smaller type such details as are of interest to the student rather than the general reader, I have attempted to adapt the book to the latter class of reader also, for many who possess a slight knowledge of geology are desirous of obtaining some insight into the geology of Lakeland, and especially to learn somewhat of the relationship of geological structure to scenery.

The photographic reproductions are from photographs taken by various people. The frontispiece and Figures 2, 12, 22, 32, 33, 39, 42 and 50 are from photographs by Messrs Abraham and Sons, Keswick; Figures 34, 35, 36, 38, 40, 41, 45 and 48 by Mr H. Bell, Ambleside; Figures 3, 4, 23, 24, 27, 31, 44 and 49 by Mr W. B. Brunskill, Windermere; Figure 14 by A. K. Coomaraswamy, Esq.; Figure 26 by Messrs Frith and Co., Reigate; Figures 25, 46 and 47 by Prof. E. J. Garwood, Sc.D., F.R.S.; and Figure 51 by R. H. Rastall, Esq., M.A. Several of these illustrations have already appeared in the volumes of Cumberland, Westmorland and North Lancashire of the Cambridge County Geographies. I am indebted to the Councils of the Geological Society and of the Geologists' Association, and also to the Editor of the *Geological Magazine* for permission to reproduce certain figures which appeared in the Journals of those Societies and in the Magazine.

I owe a special debt of gratitude to H. H. Thomas, Esq., Sc.D., of H.M. Geological Survey, for the construction of the geological map which will be found in a pocket of the cover. I am grateful to the Directors of H.M. Ordnance Survey and H.M. Geological Survey, and to the Controller of H.M. Stationery Office for leave to publish the map.

Lastly, I desire to thank R. H. Rastall, Esq., M.A., Christ's College, and W. H. Wilcockson, Esq., B.A. of Gonville and Caius College who have been so kind as to read through the proofs.

J. E. M.

CAMBRIDGE,
*December*, 1915.

# CONTENTS

# CONTENTS

# CONTENTS

## CHAPTER XII

### THE CHANGES AT THE END OF LOWER PALAEOZOIC TIMES
### (*continued*)

## CHAPTER XIII

### THE CARBONIFEROUS ROCKS

## CHAPTER XIV

### POST-CARBONIFEROUS CHANGES

## CHAPTER XV

### EVENTS IN THE DISTRICT BETWEEN THE TIME OF FORMATION OF THE DOME AND THE GLACIAL PERIOD

## CHAPTER XVI

### THE GLACIAL PERIOD: THE ICE-SHEET

## CHAPTER XVII

### THE GLACIAL PERIOD: ICE-EROSION

# CONTENTS

## CHAPTER XVIII

### THE GLACIAL PERIOD: ACCUMULATION

## CHAPTER XIX

### THE GLACIAL PERIOD: GLACIAL OVERFLOW VALLEYS. OSCILLATIONS OF THE ICE

## CHAPTER XX

### POST-GLACIAL CHANGES

# LIST OF ILLUSTRATIONS

Geological Map of the District by H. H. Thomas, Sc.D.

# CHAPTER I

## INTRODUCTION

Jonathan Otley, the Keswick Guide, must undoubtedly be regarded as the Father of Lakeland Geology. Professor Sedgwick, in a letter printed in *Life and Letters of Sedgwick* (vol. I, p. 249), says that "he was the leader in all we know of the country." The elucidation of the geology of the district dates from the publication in the *Lonsdale Magazine* for the year 1820 of a letter by Otley entitled "Remarks on the Succession of Rocks, in the District of the Lakes[1]." From that time onwards many geologists have laboured among the rocks of the district. Foremost among them was Sedgwick himself, whose writings are chiefly found among the publications of the Geological Society of London. He also gave an admirable general account of the geology in a series of letters printed in Wordsworth's *Scenery of the Lakes of England*, of which the first edition appeared in 1842, but an additional letter was published in a later (4th) edition in 1853. We may select from among the names of later workers, those of Robert Harkness, Mr J. G. Marshall, Prof. H. Alleyne Nicholson, Mr J. Clifton Ward and Mr W. T. Aveline. Mention may also be made of the value of the work of local collectors in the past, and much may yet be done by men of leisure in the district. Among these were John Ruthven, Bryce M. Wright, John Bolton and Kinsey Dover[2].

[1] *Lonsdale Magazine*, vol. I, p. 433.

[2] A list by W. Whitaker of works relating to the geology of Cumberland and Westmorland will be found in Appendix B of Mr Clifton Ward's *Geology of the Northern Part of the English Lake District*. A revised list by the same author appeared in the *Transactions of the Cumberland Association for the Advancement of Literature and Science*, Part VII, p. 13. Many papers have been written since this appeared. References to the more important of these will be given in footnotes, when treating of the subjects with which they deal.

It is my intention to describe the events that happened during the deposition of the rocks which form the Lake District, and the various changes which took place during and after their formation, as far as possible in the order in which they occurred; before doing this in detail, however, it will be well to give a general account of the geological structure of the district, and of parts of the tracts immediately bordering upon it. That part of the country to which the term "Lake District" is strictly applied is composed of rocks which are generally referred to the Ordovician and Silurian systems of the great Lower Palaeozoic group of strata, though, as will be noted in the sequel, there may be rocks of date prior to that of the two systems above mentioned.

Around these rocks is a roughly annular girdle of newer strata, partly of Carboniferous age, but partly belonging to still newer deposits of the Permian and Triassic systems. A glance at a geological map[1] will shew the general trend of the line of separation between the Lower Palaeozoic rocks and this girdle of newer strata. Beginning in the north near Caldbeck it runs a little south of west to Cockermouth, thence about south-west to Egremont, where the Carboniferous strata disappear, and Triassic rocks rest upon those of older date. From that town the line runs somewhat east of south to Millom, where Carboniferous rocks reappear. From here the line crosses the Duddon estuary and runs very irregularly in a general easterly direction past Ulverston towards Grange-over-Sands, thence in a more northerly direction to Kendal. Here the ring is broken, for the Lower Palaeozoic rocks of Lakeland are continuous with those of the Howgill Fells to the south-east, but two patches of Carboniferous rocks near Grayrigg mark the general direction of the ring. The Carboniferous rocks again set in as a continuous strip near Shap Wells, and the line is now continued

---

[1] A useful geological map for general purposes is the "Index Map" of the Geological Survey on the scale of 4 miles to an inch. The Geology of the District is represented on sheets 3 and 6 of the new series of that map. For more detailed work the Geological Survey Maps on the scale of 1 inch to a mile may be consulted. An index to these is given at the foot of the sheets of the "Index Map."

in a direction nearly north-west past Shap and the foot of Ullswater to our starting-point.

For geological purposes we may regard the Lake District as the tract of land included within the ring whose course has been roughly traced.

As Otley first shewed, the rocks within this ring, which are largely slate-rocks, may be divided into three great groups, which appear to be arranged in an arch around an axis running in a general east-north-east and west-south-west direction through Skiddaw. To the north of this axis the beds soon disappear below the unconformable Carboniferous deposits, but to the south they dip fairly steadily in a general south-south-easterly direction until they once more disappear beneath the Carboniferous rocks which form the north side of Morecambe Bay. This arrangement is shewn in the geological section

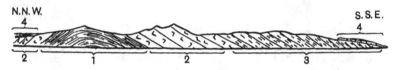

Fig. 1. Section across the District.

4  Carboniferous.      2  Borrowdale Series.
3  Upper Slates.       1  Skiddaw Slates.

drawn at right angles to the strike of the strata from a point some miles north-west of Skiddaw to the shores of the Bay near Cartmel (Fig. 1).

The lower group of strata occurs north of a line drawn from near Ullswater to the neighbourhood of Egremont, and occupies nearly all the country north of this as far as the Carboniferous rocks of the ring with the exception of a narrow strip to be noticed presently, which lies just south of the Carboniferous rocks between Penruddock and Cockermouth. Some small patches further south will be noticed subsequently. These lowest rocks have long been known as the "Skiddaw Slates." They consist largely of clayey deposits usually devoid of lime, with numerous beds of fine and sometimes coarse grit. The cleavage is generally poorly developed. The rocks were evidently laid down on an ocean-floor as sands and muds. The strata have a prevalent leaden-blue hue, at times becoming darker, and even occasionally black. The type of scenery due to the rocks of this group is illustrated in Fig. 2.

The succeeding group of rocks presents a marked contrast with these. They are almost exclusively composed of the ejectamenta shot or poured out from volcanoes. To the north they form the narrow strip above mentioned which lies south of the Carboniferous rocks of the ring. Here they are soon

Fig. 2. Melbreak and Loweswater.
Type of scenery due to Skiddaw Slates.

concealed northward by the Carboniferous strata and the succession of the Lower Palaeozoic rocks cannot be further traced upwards. To the south they extend many miles south of the line from Ullswater to Egremont, and occupy the part of the district lying between that line and another drawn from Shap Wells to Millom. In this tract is included some of the most picturesque scenery of the district—scenery which as we shall ultimately see largely owes that picturesqueness to the nature of the rocks. These rocks were described by Prof. Sedgwick as the "Green Slates and Porphyries," a good descriptive title which has largely been abandoned in favour of the term "Volcanic Series of Borrowdale," or more briefly the *"Borrowdale Series."* The latter term, as it is now in constant use, will be adopted. The view in Fig. 3 illustrates the scenery due to these rocks.

Immediately south of the rocks of the Borrowdale Series is a thin band marked by much impure limestone. It is known

Fig. 3. Windermere in winter.
The hills in the background shew type of scenery due to
rocks of the Borrowdale Series.

as the Coniston Limestone Series, and may be regarded as the base of the third division of the Lower Palaeozoic rocks, though it belongs to the Ordovician system, whereas the remainder of the third group appertains to the Silurian system.

The rocks of the third group may be briefly alluded to here as the *"Upper Slates."* Like the Skiddaw Slates they are marine sediments chiefly of clayey character, but with a larger development of gritty beds. They form on the whole rather low-lying undulating ground south of the line from Shap Wells to Millom, and extend southward to the ring of Carboniferous rocks north of Morecambe Bay. A typical example of the scenery due to these rocks is shewn in Fig. 4.

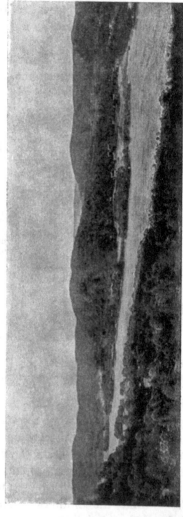

Fig. 4. Foot of Windermere.
Type of scenery due to Upper Slates.

Into the rocks of the three groups, but especially into those of the lower and middle groups, various intrusive igneous rocks have been injected which will be described in the sequel. Their positions may be noted on the geological maps.

It has been seen that the Lower Palaeozoic rocks are arranged as though folded in an arch whose axis runs through Skiddaw, and that the beds accordingly dip away from that axis, in a north-north-westerly direction on its north side, and in a south-south-easterly one to the south of it. The dips are usually at high angles, generally more than 30°, often above 60°. Far otherwise is it with the Carboniferous, Permian and Triassic strata which form the ring around these older rocks. Save where they have undergone excessive local disturbance, the dips are usually low, often between 5° and 10°, and these dips vary constantly in direction, being nearly always away from the centre of the district and at right angles to the ringed line separating these newer strata from the older rocks which form the Lake District proper. The significance of these dips will be discussed later.

It will be gathered from the above description that the Lake District proper forms a rude central dome of general upland surrounded by the ring of newer rocks, the latter as a whole giving rise to lower ground.

The table below gives the order of succession of the rocks as it now is. This has generally been regarded as the original order in which the rocks were formed, and the evidence on the

| | NAME OF DIVISION | THICKNESS | EQUIVALENTS IN WALES, ETC. |
|---|---|---|---|
| SILURIAN | Kirkby Moor Flags | to 1,500 feet | Upper Ludlow |
| | Bannisdale Slates | about 5,000 feet | Lower Ludlow |
| | Coniston Grits | to 4,000 feet | |
| | Upper Coniston Flags | about 1,500 feet | |
| | Lower Coniston Flags | 1,000 feet | Wenlock |
| | Stockdale Shales | 250 feet | Tarannon-Llandovery |
| ORDOVICIAN | Coniston Limestone | 200—1,000 feet | Ashgill Caradoc |
| | Borrowdale Series | 10,000—20,000 feet? | Llandeilo |
| | Skiddaw Slates | many thousand feet | Arenig (and earlier?)[1] |

[1] Probably also containing representatives of the Cambrian system, and perhaps even more ancient rocks.

whole seems to me to be in favour of this, though doubts have
been thrown on it; these will be considered in a later chapter.
The thicknesses assigned to the strata are only approximate.
The rocks have in many cases been much folded in detail,
and it is difficult to make accurate estimates of their original
thicknesses. The names in the third column are those of
the series as developed in Wales and the Welsh borderland,
the typical districts of the development of the Ordovician and
Silurian strata. These names are in general use.

# CHAPTER II

## LOWER PALAEOZOIC ROCKS

### A. *The Skiddaw Slates.*

The geographical distribution of the Skiddaw Slates in the
area of their main outcrop has already been noticed. In
addition to this tract other exposures of the Skiddaw Slates
are found, which are of significance as throwing light upon
movements which affected these rocks subsequently to their
deposition. There are four minor areas in the district, two
being on the north-east side, and the others on the south-west.
Of the north-easterly patches, one is developed on both sides
of the lowest bend of Ullswater, while the other, to the south-
east of this, extends from Butterwick through Bampton to a
point near the valley of Wet Sleddale. In the south-west the
mountain Black Combe is composed of Skiddaw Slates, and
another small development is found near Ireleth on the east
side of the Duddon estuary.

The general characters of the rocks have already been
stated, and in so variable a series it is of little use to enter
into points of detail. The beds are, as stated, chiefly imperfectly
cleaved slates, but there is much gritty material, which pre-
dominates in some places. For instance on the mountain
Grasmoor between Derwentwater and Crummock thick grits

are found which Mr Clifton Ward believed to be low down in the series. An interesting quartzose grit, sometimes becoming a conglomerate, was described by the same author under the name "Skiddaw Grit." It may occupy a definite horizon. This grit is seen on the north side of the Skiddaw hills striking east and west. On the west side, it is thrown northward by a great fault, and reappears east of Isell, extending past that village to Cockermouth. It is found further south to the E.N.E. of Buttermere. A similar grit is found in the south-west of the district on Latterbarrow beyond Ennerdale Lake[1].

A very important fact is the occurrence of contemporaneous volcanic rocks interstratified with the normal sediments of what is probably the highest portion of the Skiddaw Slates. Such rocks are found on Clough Head below the Borrowdale rocks of Wanthwaite Crags, in the neighbourhood of Grange in Borrowdale, on the west side of the south end of Crummock Lake, and possibly at Matterdale End near Ullswater and on the flanks of Latterbarrow to the south-west of Ennerdale. Somewhat similar rocks are associated with the Skiddaw Slates in the patch west of the village of Shap and also north of Springfield and in a tributary of Longlands Beck to the north of the foot of Bassenthwaite Lake.

The sections at Matterdale End and at Latterbarrow are difficult to interpret. There is some evidence of fault brecciation in both cases and it is possible that there are no true volcanic rocks actually inter-stratified with the Skiddaw Slates. A remarkable rock is seen in Latterbarrow Beck. It is a breccia exhibiting spheroidal structure, the brecciated fragments occurring in the bulk of the rock and in the spheroids also.

The volcanic rocks consist of thin lava-flows with beds of ash, and are of an andesitic character. It is important to notice that in each case these rocks are found near the junction of the Skiddaw Slates and the overlying Borrowdale Series.

Mr Clifton Ward, in the paper cited, attempted to make out a succession of the strata of the Skiddaw Slates on strati-graphical grounds. In rocks so highly disturbed, we know as

---

[1] Clifton Ward, "On the Physical History of the English Lake District," *Geological Magazine*, 1879, Decade II, vol. VI, pp. 50, 110.

the result of work accomplished since the death of that gifted geologist that the apparent order of succession is often untrustworthy. That being the case, the geologist must obtain what evidence he can from the included fossils of the strata. Let us therefore turn to a consideration of the organisms which have been discovered in the Skiddaw Slates.

The most abundant organisms belong to the group of graptolites. In addition to these are some trilobites and phyllocaridan crustacea, and a few brachiopods. The bulk of these fossils are undoubtedly of Arenig age, but some suggest the occurrence within the Skiddaw Slates of strata of Tremadoc age; the beds of this age may be regarded as passage beds between the Ordovician strata and the earlier Cambrian rocks.

The principal localities are naturally in the neighbourhood of Keswick, where local collectors, notably Bryce Wright, Kinsey Dover and J. Postlethwaite, have long been active, but many localities are now known at some distance from that town, and it is important that local geologists should go further afield, for many localities where fossils are known to occur have had little attention devoted to them, and no doubt numerous others await discovery.

Some of the most prolific localities lie west of Derwentwater and Bassenthwaite, as Hodgson How, Barf, Outerside, Grisedale Pike and Whinlatter road. There are several localities in the neighbourhood of Crummock Lake.

On the Skiddaw group of hills are several well-known localities, including Bassenthwaite Sandbeds, Randal Crag and Carlside Edge. A band of graptolitic rock runs along Saddleback, and what is probably the same is found in Mosedale to the east of that hill. Several localities occur in the Glenderamackin Valley and its tributaries from the Helvellyn range. Further afield from Keswick, fossils have been found at Aik Beck near the foot of Ullswater, at Thornship and Keld Becks west of Shap, and at Whicham Mill, east of Black Combe.

Little attention has been paid to the tract of Skiddaw Slates in the most northerly portion of their development, between the Caldew river and Cockermouth. Fossils have been found in this tract and as the beds here are far less affected by cleavage than those further south, a detailed examination will well repay the labour expended on it.

At present, we must regard the general succession of the Skiddaw Slates as unknown. It remains for future workers to obtain the evidence upon which alone an account of the detailed succession of the strata can be established.

But, although the general succession yet remains to be made out, we have already advanced some way towards it. Sedgwick originally suspected that these rocks contained representatives of strata developed in Wales of earlier date than Arenig times. Afterwards, as fossils of indisputable Arenig age were discovered in various localities, the Skiddaw Slates were definitely correlated with the Arenig rocks, though, as we have already seen, Clifton Ward suspected the existence of earlier strata. Among these Arenig fossils are graptolites of the genera *Tetragraptus*, *Dichograptus*, *Phyllograptus* and *Azygograptus*; trilobites of the genera *Aeglina*, and *Placoparia*, and the phyllocaridan crustacean *Caryocaris*.

It has long been known that these fossils occur in definite zones, and our latest knowledge of these zones is derived from a paper by Miss G. L. Elles, D.Sc. on "The Graptolite-Fauna of the Skiddaw Slates[1]."

The following succession is given in this paper, the names in the second column being those of the equivalent strata in Wales:

(Borrowdale Series.)

UPPER SKIDDAW SLATES:

| | | |
|---|---|---|
| (a) | Milburn Beds | Llandeilo? |
| (b) | Ellergill Beds with *Diplograptus* and *Placoparia* | Llanvirn (between Llandeilo and Arenig) |

MIDDLE SKIDDAW SLATES:

| | | |
|---|---|---|
| (a) | Upper *Tetragraptus*-beds | Upper Arenig |
| (b) | *Dichograptus*-beds[2] | Middle Arenig |
| (c) | Lower *Tetragraptus*-beds | Lower Arenig |

LOWER SKIDDAW SLATES:

| | | |
|---|---|---|
| (a) | *Bryograptus*-beds | Tremadoc |
| (b) | ? | ? Lingula Flags |

---

[1] *Quarterly Journal of the Geological Society*, 1898, vol. LIV, p. 463.

[2] It is not quite clear that the *Dichograptus*-beds are newer than the Lower *Tetragraptus*-beds.

The name Milburn Beds was given by the late Mr J. G. Goodchild to a series of Skiddaw Slates containing contemporaneous volcanic rocks which are developed among the Lower Palaeozoic rocks on the east side of the Eden Valley, to the north-east of the Lake District.    They are probably contemporaneous with the interstratified volcanic rocks of the Skiddaw Slates of the district already noticed.

Regarding the Skiddaw Slates as a whole, the fossils commonly though by no means exclusively occur in the darker deposits of the series[1].    It is to be noticed that more than one zone often occurs in a small thickness of the strata as at Barf, where zones of the Lower and Middle Skiddaw Slates are·found.

It would appear, therefore, that as is so often the case with beds containing graptolites, a frequent change in the assemblage of organisms took place during the deposition of a small thickness of strata.    But we have already seen that the Skiddaw Slates are to be measured by thousands of feet (Mr Clifton Ward suggested 10,000–12,000 feet). If this estimate be anywhere near the truth we should expect a development of faunas belonging to periods other than those of the Arenig and Tremadoc.    Either therefore the thickness of the beds has been greatly over-estimated, and the rocks of the Skiddaw Slates which have not yet yielded definite faunas are merely repetitions of the fossiliferous strata in which the organisms have been obliterated by subsequent changes, or they are of an age different from that of the fossiliferous portions.    As the newest of the latter are latest Arenig, and in some cases possibly Llandeilo, the former would be of earlier date than the latter.    The evidence, which we may briefly consider, is, on the whole, in favour of the view that we have, as Sedgwick and Ward suspected, rocks belonging to the Skiddaw Slates of pre-Tremadoc age.

Variable as are the sediments of the fossiliferous Skiddaw Slates, they possess characters in common, hard to describe, which enable us to distinguish them.    In the unfossiliferous or sparingly fossiliferous beds there are however other lithological types, which recall those of rocks of other areas of an earlier age than the Arenig.    Some of these recall rocks which are elsewhere considered to be pre-Cambrian.

The occurrence of fossils in the Skiddaw Slates at Whicham Mill, Black Combe, has already been noted[2].    Black earthy mudstones with graptolites and *Caryocaris* are here seen dipping up stream.    Higher up beds of a very different type occur.    They are now geologically above the beds at the Mill, but are almost certainly older.    Lithologically they differ from the bulk of the Skiddaw Slates, but strongly

---

[1] These dark deposits frequently contain concretions with "cone-in-cone" structure.

[2] See Smith, B., "The Glaciation of the Black Combe District," *Quarterly Journal of the Geological Society* (1912), vol. LXVIII, p. 406.

recall in appearance the beds near Ingleton which have been claimed as pre-Cambrian. Still higher up the beck and forming the mass of Black Combe are deposits similar in appearance to the more normal types of Skiddaw Slates.

Rocks which recall types of pre-Cambrian sediment are also found about Latterbarrow near Ennerdale, and among the uplands at the back of Skiddaw.

In this last tract are some sediments of a different type, which I have not seen elsewhere among the Skiddaw Slates. They are developed in Little Wiley Gill and Hause Gill near Great Cockup, on which Clifton Ward's "Skiddaw Grit" is exposed. The beds in the gills consist of cleaved shales. Some are black banded shales, others red and others again green. In the black and green shales I found fragments of large trilobites, but nothing enabling one to distinguish even the genus. The beds somewhat resemble Cambrian beds of other districts and would well repay further examination, especially as, by means of them, the question of the age of the Skiddaw Grit may be ultimately solved.

From what has been said, it is obvious that much detailed work is wanted among the complex series of rocks of the Skiddaw Slates.

There are various lists of fossils from these slates. A complete list of the graptolites hitherto discovered will be found in Miss Elles' paper. Other organisms are recorded in Clifton Ward's memoir on the geology of the northern part of the English Lake District. A paper on some trilobites from the Skiddaw Slates by Postlethwaite and Goodchild is published in the *Proceedings of the Geologists' Association*, vol. IX, and there is a notice of some brachiopods and molluscs in a little work by Postlethwaite on *The Geology of the English Lake District* (Keswick, 1897). In these two works figures are given, and some new specific names are proposed, but the material is very imperfect. The occurrence of *Placoparia* at Outerside was recorded by Miss Elles and Dr Cowper Reed in the *Geological Magazine* for 1898 (pp. 141 and 240).

# CHAPTER III

## LOWER PALAEOZOIC ROCKS

### B. *The Borrowdale Series.*

The distribution of the main development of these rocks north and south of the Skiddaw Slates has already been noted. In addition, there is an outlying patch projecting through the Carboniferous rocks of Greystoke Park and another on the east side of the Duddon estuary.

The relationship of these rocks to the Skiddaw Slates will be more fully considered hereafter. It is perfectly clear that they are now resting upon Skiddaw Slates at the junction of the two series in most places, but it is equally clear that the junction is usually a faulted one, though the inclination of the fault is not far removed from the horizontal plane.

In one place however there appears to be evidence of an unfaulted passage from the Skiddaw Slates to the rocks of the Borrowdale Series. Its significance was first grasped by Prof. H. A. Nicholson. The section occurs in the cliff behind the Hollows Farm, south-west of Grange-in-Borrowdale. The junction is clearly seen for a distance of several yards, and as hand specimens can be secured in which the rocks of the two series are actually welded together, the unfaulted character of the junction seems to be beyond doubt.

The highest beds of the Skiddaw Slates are here greenish micaceous mudstones, from which a specimen of *Lingula* was obtained within two inches of the plane of junction. The lowest member of the Borrowdale Series is a dark green ash, which extends up the cliff for a considerable distance. Behind the lower part of the cliff on which this section occurs is a depression separating it from the main mass of cliff extending up towards Maiden Moor. There is little doubt that the fault which in an easterly and westerly direction separates the Skiddaw Slates from the Borrowdale volcanic rocks, leaves that junction behind the Hollows Farm, and occurs locally in the Borrowdale rocks at some distance above the junction with the Skiddaw Slates.

Mr J. F. N. Green maintains that the Borrowdale rocks rest conformably on the Skiddaw Slates near the Duddon estuary and between Haweswater and Shap (see *Geol. Mag.* Decade VI, vol. II (1915), p. 189). In a paper published since this was written, Green asserts the occurrence of a similar conformity in the eastern part of the district (*Proc. Geol. Assoc.* vol. XXVI (1915), p. 195).

The rocks of the Borrowdale Series with some local exceptions to be noticed shortly are exclusively volcanic: they consist of material poured forth from volcanic vents as lavaflows, or shot out in a fragmental condition to give rise to volcanic ashes and agglomerates. Apart from the exceptions mentioned below, they appear to be absolutely devoid of fragmental material which was derived from the ordinary erosion of the land. No fossils have been found in them to give indication as to the condition under which they were formed, whether by terrestrial or submarine volcanoes. Mr Clifton Ward maintained their terrestrial origin but the evidence on this point is not convincing.

Before describing the characters of the volcanic rocks, further mention may be made of the possible occurrence of true sediments associated with the volcanic materials.

In Cat Gill, which descends from Falcon Crag near Keswick, above the lowest lava of the Borrowdale rocks is a series of ashes and breccias, alternating with thin bands of shaley grey mudstone which are obviously sedimentary though greatly crushed. They certainly appear to be interstratified with the volcanic rocks, though it is just possible that they are masses of Skiddaw Slates dragged in along minor fault-planes above the great junction fault.

The only other example of an apparent sedimentary rock in this series is in a small slate quarry situated near the path through Hole Rake, a col separating the Church Beck Valley, Coniston, from a stream draining into Tilberthwaite. There a few feet of grey crystalline limestone occur in a highly folded condition among the volcanic slates. The limestone appears to be a true sediment. It may possibly be a mass of the overlying Coniston Limestone Series folded in among the volcanic rocks, but its mode of occurrence in the ashes suggests that it is really an interstratified sediment, for the folding and faulting required to bring it to its present position on the other supposition would be of an extraordinarily complex character.

The general nature of the rocks of the series is well described in Prof. Sedgwick's term "green slates and porphyries" though it must be remembered that the slaty structure was superinduced as the result of subsequent change. The prevalent green hue is very marked, and forms a strong contrast with the Skiddaw Slates below and the upper slates above, where grey, blue and black are the predominant hues.

Many of the lavas are porphyritic, the principal large crystals being in some cases felspar, in others augite, and sometimes both occur in the same rock. The slates are very largely the result of cleavage impressed upon fine volcanic ashes, which must have been showered forth as fine dust, but where the agent producing cleavage has acted with exceptional intensity coarse ashes and even breccias and lavas have been converted into slates.

The lavas vary in thickness, some being only a few feet thick, while others are of much greater thickness. Many of the lavas shew a vesicular structure; in the thin flows this structure extends throughout, while in the thicker masses it is confined to the surfaces. These vesicles have been subsequently filled with various minerals as chalcedony, agate, calcite and chlorite, converting the lavas into amygdaloidal rocks.

A characteristic feature of these rocks is the abundance of garnets. Many of them are of microscopic dimensions, while others are as large as a pea. They occur in lavas and ashes alike, and are particularly abundant and well formed around the central Scawfell hills. Good examples are procurable just below Sty Head Tarn; on Illgill Head (the hill above the screes at Wastwater); on Gunson Knott (an eminence of Crinkle Crags); and around Haweswater on the top of Kidsty Pike and in several places on the hills between that lake and the valley of Swindale.

The lavas and ashes vary in composition. Among the lavas we get varying percentages of silica so that the rocks range from basic andesites to acid rhyolites. Whatever be the composition of the lavas, the fragments in the breccias are often of an acid character, as though a scum of lighter material rose to the top of the liquid rock filling the vents and was showered out again and again as the result of explosions.

The fragments of the breccias vary much in size. Some are so small that it is equally correct to speak of the rock as coarse ash or fine breccia. Many are an inch or more in diameter, and in some cases we find fragments which measure many inches across.

The lowest rocks of the series are essentially andesites similar to those which we noted as occurring in the Skiddaw Slates, while the highest division is largely composed of rocks of a rhyolitic character. The significance of this will be discussed in a later chapter.

In many parts of the district the rocks have undergone movements of so profound a nature after their formation that the separation of lavas from ashes is a matter of considerable difficulty. Elsewhere however, lavas and ashes can be

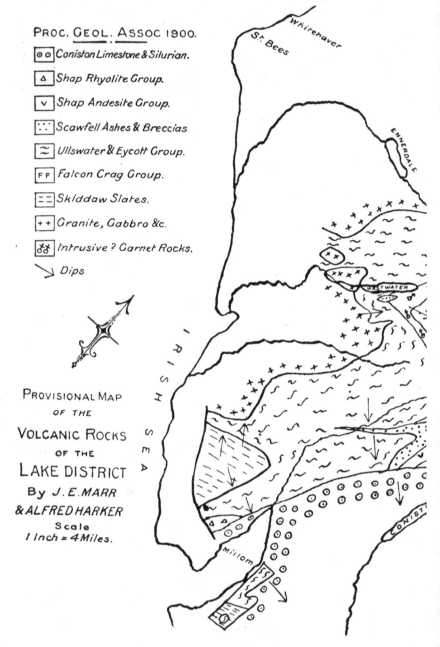

PROC. GEOL. ASSOC 1900.

⊙⊙ Coniston Limestone & Silurian.

△ Shap Rhyolite Group.

∨ Shap Andesite Group.

∴ Scawfell Ashes & Breccias

≈ Ullswater & Eycott Group.

F F Falcon Crag Group.

= = Skiddaw Slates.

+ + Granite, Gabbro &c.

⚭ Intrusive ? Garnet Rocks.

↘ Dips

PROVISIONAL MAP
OF THE
VOLCANIC ROCKS
OF THE
LAKE DISTRICT
By J. E. MARR
& ALFRED HARKER
Scale
1 Inch = 4 Miles.

Whitehaven

St Bees

ENNERDALE

WASTWATER

IRISH SEA

Millom

CONIST

Fig.

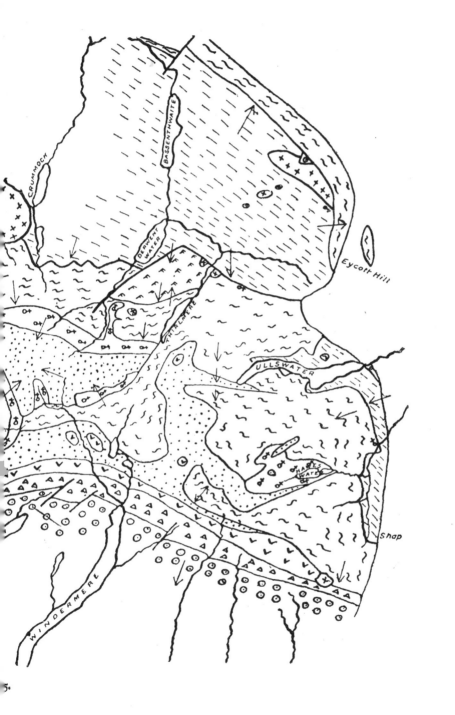

CRUMMOCK

BASSENTHWAITE

DERWENT WATER

THIRLMERE

Eycott Hill

ULLSWATER

HAWES WATER

Shap

WINDERMERE

distinguished with ease even from a distance, as the lavas tend to give rise to well-marked cliffs, while the ashes, which are less durable, are marked by sloping ground. Such is the case among the group of hills separating Derwentwater from the Vale of St John. A spectator standing on the west shores of Derwentwater will see terrace after terrace rising from just above the shores of the lake over Falcon Crag right up to the summit of Bleaberry Fell. The same appearance is well seen in the volcanic rocks which are found north of the Skiddaw Slates. These are usually spoken of as the Eycott Series, from Eycott Hill north of Troutbeck station and there the terraced character is strikingly exhibited.

As the rocks of the Borrowdale Series immediately succeed the highest Skiddaw Slates of late Arenig or possibly early Llandeilo age, and as the volcanic rocks are themselves succeeded by marine strata with Caradocian fossils, the series, if the present succession be the original one, must belong as generally stated to the Llandeilo division.

Various estimates have been made as to the thickness of the series, and as one would naturally expect, in the case of rocks which have undergone so much disturbance, the difference between the greatest and smallest thickness assigned to them is considerable.

In the table on p. 7 a thickness of from ten thousand to twenty thousand feet is mentioned. In the light of recent researches, the higher estimate is probably nearer the truth, and even this may be too low[1].

In a paper by Dr Alfred Harker and myself[2], a classification of the various members of the Borrowdale Series is attempted, and in the same volume a rough map shewing the general distribution of the sub-divisions is inserted facing p. 537. A section was also drawn across the series which was subsequently reproduced with corrections in the "Jubilee Volume" of the Association (1910), p. 631. The map and section are here reproduced as Figs. 5 and 6.

It must be noticed that the map is confessedly generalised and not founded on a detailed survey. Important modifications in detail will

---

[1] The reader will find an admirable general account of these rocks by Sir Archibald Geikie in his *Ancient Volcanoes of Great Britain*, vol. I, pp. 227–238.

[2] *Proceedings of the Geologists' Association*, 1900, vol. XVI, p. 449.

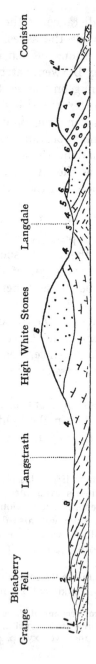

Fig. 6.  Volcanic Series from the south end of Derwentwater to the north end of Coniston.

8  Coniston Limestone.

7  Shap Rhyolitic Group.

6  Shap Andesitic Group.

5  Scawfell Ash and Breccia Group.

4  Sty Head Garnetiferous Group.

3  Ullswater Basic Andesite Group.

2  Falcon Crag and Bleaberry Fell Group.

1  Skiddaw Slates.

L'  Fault between Skiddaw Slates and Borrowdale Series.

L"  Fault between Borrowdale Series and Coniston Limestone Series.

no doubt result from such a survey, but I am confident that the divisions and their general distribution on the ground will be found correct.

The following table gives these divisions:

*Coniston Limestone Series.*

Shap rhyolitic group.
Shap andesitic group.
Borrowdale    Scawfell ash and breccia group.
Series    Sty Head garnetiferous group.
Ullswater basic andesite group = Eycott group.
Falcon Crag group.

Skiddaw Slates.

It will be convenient to consider these divisions in the order of their antiquity, but before doing so reference may be made to the present lie of the rocks as shewn in the map and section. On the whole they dip in a general south-south-easterly direction in the central part of the district, but folding occurs. A gentle syncline occupies the position of Bleaberry Fell with a gentle anticline to the north about Watendlath. A much more important syncline has its centre in the Scawfell group of hills, and it is succeeded to the south by an anticline crossing the Great Langdale Valley. Other folds will be noticed on an examination of the map.

(a) *Falcon Crag group.* The lowest and presumably oldest rocks of the series have a limited areal distribution: nevertheless it is this group in which occurs the section described by Mr Clifton Ward as constituting a typical section of lavas and ashes[1].

This, the Falcon Crag group, extends in an east and west direction from the western side of the Vale of St John to the east side of Borrowdale, and north and south from the junction with the Skiddaw Slates to the neighbourhood of Watendlath Tarn.

The lowest rock seen is a massive purple breccia, forming an isolated crag, just above the road below Falcon Crag; it is also seen in road-cuttings near this. We shall see reasons for supposing that this rock is a fault-breccia. It contains abundant fragments of Skiddaw Slates and of Borrowdale rocks. Between this and the summit of Falcon Crag, Ward describes twelve lavas with intervening ash beds: there are also minor lavas of a vesicular character. It is interesting to note that the purple breccia was inserted by William Smith in his Geological Map of Cumberland which appeared in 1824.

The ashes above the lowest lava contain the seams of sediment already referred to as developed in Cat Gill. An ash between the two

[1] *Geological Survey Memoir,* "The Geology of the Northern Part of the English Lake District," pp. 13–19.

lowest well-marked lavas contains abundant small garnets. The highest lava is also marked by an abundance of this mineral in places.

The lavas of this division are pyroxene andesites containing 59–61 per cent. of silica. The pyroxene may be hypersthene, augite or both occurring together. Porphyritic plagioclase felspars are often present. "To the eye the rocks are pale greenish to dark grey, with a compact ground-mass, usually containing scattered minute porphyritic felspars, or more rarely abundant crystals up to ¼ inch in length. Many of the flows are amygdaloidal. There are rare flows of more basic lavas besides some tuffs[1]."

From the great thickness of the rocks of this group we can hardly suppose that their distribution at the surface had so limited an east and west extension as is now shewn. There is reason to suppose that to the east and west of the present outcrop, the rocks of the Falcon Crag group are faulted out, except behind Hollows Farm, so that those of the next group are brought into direct contact with the Skiddaw Slates.

(*b*)　*Ullswater group = Eycott Hill group.* The rocks of this group have a much wider distribution than have the Falcon Crag rocks. An inspection of the map will shew their extensive development on either side of Ullswater, extending southward towards Shap. They occupy the lower part of the Thirlmere Valley, and extend through Dunmail Raise to the south of the central syncline. North of that syncline they are traceable westward from Thirlmere to Borrowdale, but on the east side of the latter valley are poorly represented, probably owing to concealment of a large portion by faulting. On the west side of the valley they again appear in force, and a fine sequence may be traced from the junction of the Skiddaws near Maiden Moor, by passing southward along the slope past Eel Crag and Dale Head to Honister. At Honister beds of fine ashes in this group are well cleaved, and form the well-known Honister Slates. The slate band is traceable from Honister to Borrowdale, and also crosses that valley, but the band of intense cleavage appears here to transgress from one horizon to another, and even eventually to affect the rocks of the Falcon Crag group. On the south side of the central syncline the Ullswater group is traceable from the Grasmere Valley across the Great and Little Langdales into the Duddon Valley. Here the outcrop again widens, and the rocks occupy a large area between Wastwater on the north and Black Combe on the south. Rocks referred to this group are traceable on the east side of Black Combe, and also on the east side of the Duddon Valley. In a paper by Mr J. F. N. Green entitled "The Older Palaeozoic Succession of the Duddon Estuary" (issued by the Author) detailed maps of the

---

[1] Notes on the principal varieties of the lavas of the Borrowdale Series by Dr Alfred Harker are published in the *Naturalist* for 1891, and in the *Proceedings of the Yorkshire Geological and Polytechnic Society* for 1902.

succession in the tract between Duddon Bridge and Millom and in the Ireleth tract are given, and the succession of the rocks is described. If we eliminate the intrusive rocks in this tract, the thickness of the volcanic rocks which I would refer to the Ullswater group is here small— between 500 and 600 feet. A map of the rocks on the east side of the district, according to Mr Green's interpretation, accompanies the paper cited on p. 14.

The rocks of this group generally resemble those of the underlying group as seen in the field. They also consist of alternations of lava, breccias and ashes, and the field geologist would see no reason for separating the two divisions. The lavas as a whole are more massive than those of the Falcon Crag group, and phenocrysts of felspar and augite are usually more conspicuous. The main difference between the two groups is one of composition, and it is of importance. The bulk of the lavas of the group under consideration are more basic than those of the underlying group, containing 51 to 53 per cent. of silica instead of 59 to 61 per cent. They are *basic* andesites. It is true that some intermediate andesites are found, just as basic andesites occur sparsely among the prevalent *intermediate* andesites of the Falcon Crag group, but they are rare[1].

We may now notice some interesting features in the characters of these rocks. A purple breccia is often found at the base, where the rocks of this group are brought against the Skiddaw Slates, and there is evidence that this, like the similar breccia at the base of the Falcon Crag group, is a fault-breccia.

The occurrence of large phenocrysts is of importance for a reason to be noted presently. These are found in some of the rocks on the slope west of Borrowdale, but are best developed in the neighbourhood of Haweswater, as on the south side of Riggindale, but especially in the tract to the north of the foot of the lake.

A remarkable nodular lava is found in this series, crossing Fordingdale Beck, Haweswater, and extending to the top of Kidsty Pike, where it is well shewn just below the summit on the southern face. Its outcrop is noticed by Dakyns[2]. It no doubt extends much further than is at present known, and as the above writer indicates will prove a useful horizon for purposes of detailed mapping.

It has been seen that garnet-bearing rocks occur in the Falcon Crag group: they are also found in the group under consideration for instance on the east side of Borrowdale near Rosthwaite. It is probable that some of the rocks indicated as belonging to the Sty Head

[1] See Harker, *Naturalist, loc. cit.* Also the same, "Chemical Notes on Lake District Rocks," *Naturalist,* 1899.

[2] *Geological Survey Memoir,* "Geology of the Country between Appleby, Ullswater and Haweswater," pp. 17, 18.

garnetiferous rocks on the map really belong to the present group, as for instance the tongue shewn between Borrowdale and Watendlath, and others in the neighbourhood of Haweswater.

Let us turn northward, to the belt of volcanic rocks occurring north of the Skiddaw Slates. These formed the subject of a special paper by Clifton Ward "On the...Lavas of Eycott Hill, Cumberland[1]" and notes upon the rocks by Dr Harker will be found in the papers in the *Proceedings of the Yorkshire Geological and Polytechnic Society* and the *Naturalist* to which references have been given. They have long been known as the Eycott Hill rocks, and their age has been a matter of doubt. When the Falcon Crag rocks were supposed to be typical representatives of the Borrowdale group as a whole, the difference in the chemical composition of these and of the Eycott rocks, which latter contain 51–53 per cent. of silica, was felt to be a difficulty. The Eycott rocks have even been claimed as of possible Devonian age. The subsequent discovery of rocks in the main development of the Borrowdale Series, to wit those of the Ullswater group above described, removes the difficulty and the similarity of the rocks with large phenocrysts of felspar around Haweswater, which precisely resemble some of the lavas of the Eycott Hill group, is strong evidence that the two groups are really contemporaneous: they are accordingly treated together here. As the Eycott Hill rocks have not been affected by the cleavage which has produced such marked effects on the rocks to the south, they display their original characters in an admirable manner.

They consist of the usual alternations of lava, ash and breccia, many of the thinner lavas being vesicular throughout. Two noticeable rocks are found. The first is the lowest lava, which is traceable for a long distance north-west and west from Eycott Hill. It is marked by the extraordinary size of the felspar phenocrysts, which in some cases are two inches in diameter. The other rock is a little higher in the series. Its phenocrysts are smaller than those of the rock last mentioned, and generally there is nothing noteworthy to distinguish it from the other porphyritic lavas of the group. On Greenah Crag, however, to the south of Eycott Hill it is stained red, and in this condition forms one of the most beautiful rocks of the district.

The rocks of Greystoke Park have not been studied in detail. They form an inlier among the Carboniferous rocks, appearing at the surface owing to a fault. They are dipping away from the rocks of the Eycott tract, and may be of newer age than those, but the effect of the fault, on the other hand, may merely be to cause a repetition of the Eycott rocks.

(c) *Sty Head garnetiferous group.* The rocks of this group are admirably displayed when ascending to the top of Sty Head Pass from Stockley Bridge.

---

[1] J. Clifton Ward, *Monthly Microscopical Journal*, 1877, p. 239.

They occur in many places between the rocks of the Ullswater group and those of the succeeding Scawfell group. Their distribution as indicated on the map is no doubt in many respects incorrect. As already noted, some of the rocks so indicated may belong to the earlier group, and some may be intrusive. On the other hand they probably occur in places not shewn on the map, as for instance below the rocks of the Scawfell group on the Helvellyn range. On Sty Head and all around the central Scawfell group of fells the rocks have a remarkable flinty appearance, due to subsequent alteration. I suspect that this alteration has not affected the rocks of Helvellyn, and that certain rocks on that range which do not shew the flinty character, but which nevertheless present other characters resembling those of the rocks of Sty Head, really belong, as their position indicates, to the Sty Head group. These rocks are well shewn on the western slopes of the Helvellyn range.

They are spoken of as "garnetiferous rocks" on account of the great abundance of garnets in the rocks of certain tracts. The mineral occurs alike in lavas, fine ashes and breccias. The rocks of this group have been described in detail in a paper by the late Mr E. E. Walker[1].

The rocks of this division are as a whole of a much more acid character than those of the underlying group. Walker gives two partial analyses of these rocks which respectively give silica percentages of about 69 and 76.

The main feature of these rocks is a remarkable "streaky" structure, described in detail by Walker. It is not confined to the rocks of this group, but is certainly a dominant feature of it. It occurs both in lavas and fragmental rocks, and has therefore not always originated in the same way. Walker gives four methods of production, all of which he believes to be exhibited by the rocks of this group. They are:

(1) By infiltration of quartz, calcite, chlorite, epidote and ilmenite along definite planes.

(2) As the result of lamination of fragmental rocks.

(3) As the result of flow of igneous material.

(4) As the result of dynamic action on included fragments.

Where the rocks are altered it is often difficult to discover the manner in which the streaky structure has been developed, and indeed to know whether we are dealing with a lava or an altered ash.

The origin of the garnets is also a difficult question which cannot be dealt with here at length. A study of the rocks led Walker to conclude that the evidence was in favour of the occurrence of garnets as original minerals, and not as the result of subsequent alteration of the

---

[1] "The garnet-bearing and associated Rocks of the Borrowdale Volcanic Series," *Quarterly Journal of the Geological Society*, 1904, vol. LX, p. 70.

rock. Another mineral of interest occurring in a lava of this group is cordierite[1].

(*d*)  *Scawfell Ash and Breccia group = Tilberthwaite Slates.* The rocks of this group have a wide distribution in the centre of the great syncline which extends through the Scawfell hills eastward through the Helvellyn and High Street ranges. Along this part of the tract they are marked as a general rule by their flinty character, which is the result of subsequent changes. To the south of this development we meet with a remarkable band of slate which, first seen in the eastern part of the district in the valley of Mosedale, is traceable westward and then south-westward past Long Sleddale, Kentmere, Troutbeck, Langdale, Little Langdale, Tilberthwaite, Coniston Old Man, Walney Scar, and then, with a sharp southerly bend to a point where it abuts against the Coniston Limestone. Along the whole of this line of outcrop the slate band is readily traceable by the numerous slate-quarries.

We shall presently give reasons for concluding that the rocks of the Scawfell syncline and those of this slate band belong to the same group, though they are usually so different in appearance.

Beginning with a description of the rocks of the central syncline around Scawfell, we may first notice the prevalence of pyroclastic material changing from very coarse breccia through insensible gradations to the very finest volcanic dust. Lavas are no doubt present in places, but the dominant feature is the excess of the fragmental rocks. They are also characterised by their even bedding, and the finest ashes in particular often possess a very regular lamination. These banded ashes are quite common in any of the divisions of the Borrowdale Series, but nowhere do they attain the importance which they assume in the group under consideration. The finer ashes are often concretionary. A remarkable example occurs on the Eskdale side of the top of Esk Hause. Other interesting concretionary rocks with much epidote are developed near the head of Greenup Gill, and at the foot of the Stake Pass in the Langstrath Valley.

On the Scawfell hills the apparent thickness of these ashes is very great. They are found dipping at a gentle angle towards the centre of the syncline on "Esk Pike," a name given by Ward to a previously nameless hill between Scawfell Pike and Bowfell. Viewed from a distance the apparent major bedding planes seem to be oblique to the minor and undoubtedly true planes of bedding, and this and other reasons into which we need not enter suggest that this apparent thickness is due to repetition along fault-planes which are nearly horizontal. The beds of the Tilberthwaite slate bands have a thickness which is much less than the apparent thickness of these rocks around Scawfell.

[1] See Harker, "A Cordierite-bearing Lava from the Lake District, *Geological Magazine*, 1906, Decade v, vol. III, p. 176.

The composition of these rocks varies considerably, but is in accord with the supposition that they have been derived from a magma of an andesitic character sometimes approaching the rhyolites. The following silica percentages are quoted by Harker in his paper "Chemical Notes on Lake District Rocks[1]": Slight Side near Eskdale 68·421, Grasmere 61·75, Tilberthwaite Quarries 61·25 (the two last are from the Tilberthwaite slate band).

Turning now to the slate band to the south, we find that all the lithological characters of the rocks which are not due to subsequent alteration are similar to those of the Scawfell ashes. The same variation from very fine to coarse ash is found, and the same large proportion of finely laminated volcanic dust. Some of these ashes are locally known as "bird's-eye Slates." They contain flattened elliptical portions with a light outer ring surrounding a dark green centre. These are usually regarded as concretions.

Both the Scawfell ashes and the Tilberthwaite slates lie above the Sty Head group and are succeeded by the Shap andesite group. They are as we have seen similar in character, save that they have undergone different types of alteration, the former having the flinty character probably produced by heat, the latter the slaty cleavage due to lateral pressure.

In some places the two types seem to come together where the syncline joins the slate band, as on Red Screes. The most convincing evidence however is found near Walney Scar between Coniston and the Duddon Valley. At this place the flinty type of alteration is locally developed on a fairly large scale, and the rocks presenting this character are seen actually passing into those with the normal features of the rocks of the slate-band within a very short lateral distance.

Remarkable effects of lateral pressure in producing folding, faulting and brecciation of these rocks will be noticed in a later chapter when describing more fully the changes that were brought about in the rocks of the Lower Palaeozoic group subsequently to their formation.

Dr W. Maynard Hutchings has made a detailed examination of the microscopic characters of some of the slates of the Tilberthwaite slate bands[2]. He finds that minute garnets, quite invisible to the naked eye, are abundant in the slates. This is important. We have seen that garnets of fairly large size are abundant, and may well suspect that even when invisible, garnets do contribute to the formation of the rocks of other groups of the Borrowdale Series as well as to those of this slate band. It may be noted that no garnets have been recorded in

---

[1] *Naturalist*, February and May, 1899.

[2] Hutchings, W. M., "Notes on the Ash-Slates and other Rocks of the Lake District," *Geological Magazine*, 1892, pp. 145 and 218.

unaltered rocks of the Eycott Hill group. They may occur in the microscopic condition.

(*e*) *Shap Andesitic group.* The rocks of this group are of no great thickness. They form a well-defined band between the preceding and succeeding rocks, being developed from the Shap granite to the moorland west of Coniston Lake. They repeat in their main features the characters and composition of the Falcon Crag group. Like them they are intermediate andesites. Harker quotes the case of an andesite between Wasdale Pike and Great Yarlside, west of the Shap granite, with 59·95 per cent. of silica. The lavas are generally thin and often vesicular throughout.

(*f*) *Shap Rhyolitic group.* These rocks occur immediately south of the preceding, from the Shap moorlands to those of Coniston. They consist as usual of lava-flows, ashes and breccias. The lavas are usually pink, and the pink colour of the fragments in the breccias is a noteworthy feature. The rhyolitic lavas are best seen towards the eastern end of the district, and are specially well developed on the east side of the Shap granite. They often present a lamellar structure due to flow; they are seldom porphyritic or vesicular, and when vesicles occur they are usually of microscopic size. A rhyolite from near the Shap granite has a silica percentage of 79·65.

On the west side of the Yewdale Valley, north of Coniston Lake, an exceptional thickness of rhyolitic breccia occurs in this group, and gives rise to the prominent cliffs of Yewdale Crags. It was specially mentioned by Clifton Ward as the Yewdale Breccia.

An interesting question arises as to whether the volcanic rocks as a whole give any indication of changes in the composition of the rocks as the result of differentiation of the magma from which they were derived. Taking a broad view, there does appear to be evidence of an earlier eruption of rocks of intermediate composition (the Falcon Crag group), followed by more basic rocks on a large scale (Ullswater group): this was followed by a minor eruption of intermediate rocks (Shap andesites) and the last phase was marked by the emission of acid rocks (Shap rhyolites).

It is true that rocks of rhyolitic character may contribute in a subordinate degree to any of the divisions and that fragmental rocks of acid character were certainly shot out at various times, but, as has been seen, this may be simply due to local differentiation in the actual pipe or fissure, and not to any changes in the deeper seated magma.

# CHAPTER IV

## INTRUSIVE IGNEOUS ROCKS CONNECTED WITH THE BORROWDALE SERIES

A large number of intrusive igneous rocks have been injected into the sedimentary and contemporaneous volcanic rocks of the Lake District. These are not all of one period, and the determination of their age is in some cases a matter of difficulty. It is clear that they are of later date than the rocks into which they are actually intruded; the difficulty is to discover how much later. Various kinds of evidence are available. Pebbles of the intrusive rocks may be found in newer rocks, shewing that the intrusion occurred before those rocks were formed. The intrusive rocks when associated with volcanic rocks may shew such similarities of composition when compared with the latter as to indicate genetic relationship; in this case the intrusive rocks are newer than the volcanic rocks into which they have been forced, but contemporaneous with the volcanic group as a whole, having been forced into older rocks beneath the surface, while contemporaneous volcanic material was in course of accumulation at the surface. A third test of great value in the Lake District is connected with the effects of earth-movements on the rocks. We shall ultimately see that at the end of Lower Palaeozoic times the rocks of the district were affected by great movements. These have impressed their effects upon some of the intrusive igneous rocks, which were therefore consolidated before the movements occurred, while another set not affected by them is obviously of later date. Accordingly we can divide the intrusive rocks of the district into two groups, an older and a newer[1].

Many of the rocks of the older group are clearly associated with the volcanic rocks of the Borrowdale Series, and the composition of some of the sills and dykes is so closely akin

[1] See Harker, *Proceedings of the Yorkshire Geological and Polytechnic Society*, 1902, vol. XIV, p. 487.

to that of the lavas that their genetic relationship is indisputable. Others, as the result of differentiation, differ in composition from the lavas. A large number of these contain garnets of the same type as those found in the contemporaneous volcanic rocks, and the presence of these garnets probably shews their relationship. Again, these igneous rocks of the older group are affected by the earth-movements mentioned above. Rocks of Silurian age would also be affected by such movements: as however the evidence at our disposal indicates quiescence of the district in Silurian times, so far as any igneous activity was concerned, the probability is that the rocks which bear the impress of these earth-movements are of Ordovician date.

The intrusive rocks which can definitely or with probability be referred to the period of formation of the volcanic rocks of the Borrowdale Series were in many cases forced among the rocks of that series. Others were injected along the plane of junction between the Borrowdale rocks and the Skiddaw Slates, and others again among the rocks of the Skiddaw Slates.

The rocks vary in composition, from very basic olivine-bearing rocks to granites and quartz-porphyries. They occur in various ways. Some are large masses of laccolithic character forced along important divisional planes, as the Ennerdale and Buttermere granophyre, and the St John's Vale microgranite. Others occur as minor laccoliths like the masses in the Langstrath and Seathwaite Valleys, others as intrusive sheets or sills, and a vast number as wall-like masses or dykes with a more or less vertical position. A very network of these dykes pierces the volcanic rocks in the Scawfell group of fells.

Among the more interesting rocks we may notice the olivine-diorite or picrite of Little Knott and Great Cockup, intrusive in the Skiddaw Slates in the northern portion of the Skiddaw group of fells, consisting largely of hornblende with an altered product of olivine: this is the most basic of the intrusive rocks. Less basic are the dolerites or diabases without olivine, consisting of plagioclase felspars and augite usually much altered by decomposition. These are abundant in the Skiddaw Slates and Borrowdale Series. The rock of Castle Head near Keswick is an example.

Of the acid rocks the largest mass is the Ennerdale and Buttermere granophyre, forced along the junction of the Skiddaw Slates and extending from Scale Force near Crummock to the foot of Wastwater. It is usually a pink rock composed of felspar, quartz, and chloritic substance due to the alteration of a basic mineral. A somewhat similar rock well seen at Blea Crag in the Langstrath Valley contains garnets. A well-known dyke—the Armboth and Helvellyn dyke—also contains crystals of that mineral. It is seen on one or two places on the east side of Armboth Fell, and what is probably an extension of it occurs on Helvellyn. The laccolithic mass of igneous rock which is quarried at Threlkeld is somewhat similar, but of coarser grain, being a microgranite. It is usually known as the St John's Vale microgranite. It also contains garnets in places. Lastly we find numerous dykes of the quartz-porphyry class, especially towards the centre of the district. Some of a less acid character may be spoken of as quartz-porphyrites.

The study of the metamorphism of the rocks into which these igneous rocks have been injected is a matter of interest. The metamorphism produced by the igneous rocks of the older group has not been studied with the attention to detail which has been paid in the case of some of the rocks of the newer group. On the whole it may be stated that the amount of alteration produced by these older rocks is much less than that brought about in the case of some of the newer group.

It is desirable that further details should be given about some of the more important of these intrusive rocks, with references to papers in which the reader may find fuller accounts.

The picrite of Little Knott was described by Professor Bonney[1] and some notes will be found in Harker's paper cited above. A detailed examination in the field will doubtless repay the undertaking. It is probable that rocks of a more acid character which occur in association with this rock are in genetic connexion with it, and that we are dealing here with a case of magmatic differentiation. Notes on the Great Cockup rock are given by Prof. Bonney in a paper by J. Postlethwaite[2].

Of the dolerites Harker[3] writes as follows: "They form dykes and sills, and exceptionally a small boss (e.g. Castle Head, Keswick). They

[1] Bonney, T. G., *Quarterly Journal of the Geological Society*, 1885, vol. XLI, p. 511.
[2] Postlethwaite, *ibid.*, 1892, vol. XLVIII, p. 508.          [3] *Loc. cit.*

are mostly ordinary dolerites without olivine, often considerably altered with production of chlorite, etc. Usually they appear in the field as dull, dark-coloured, medium to fine-grained rocks, without porphyritic elements. The Castle Head rock, however, has porphyritic augites, and is in places micaceous."

There is little doubt that many intrusions of intermediate composition occur. It is extremely difficult, when these are found among the rocks of the Borrowdale Series, to distinguish them from true lava-flows. Many of the more massive fairly crystalline rocks of that series which have been mapped as lavas may turn out to be intrusive. Mr J. F. N. Green writing of the Millom and Ireleth tracts states that "there can be no doubt that many of the pyroxene-andesites are intrusive, their upper edges transgressing and hardening the super-incumbent beds and sometimes enclosing masses of them." He further points out that some of them are actually intrusive into the Skiddaw Slates.

Of the quartz-porphyries and quartz-porphyrites Harker writes[1]: "these are included together because it is scarcely possible to draw any true line between the truly acid and the sub-acid rocks. Dykes and sills are found at numerous localities, and occasionally more irregular intrusions (e.g. Wansfell). The rocks are in no wise remarkable."

Of the microgranites, the St John's Vale rock is the best example. "This is a fine-textured grey rock with small porphyritic felspars scattered through it. These are oligoclase. Flakes of biotite are also present, and the general mass of the rock is a microcrystalline aggregate of quartz and felspars." A specimen of the rock gave a silica percentage of 67·18.

Some interesting features are noticeable among the rocks of the Threlkeld quarries. Masses of Skiddaw Slates (often of large size) and of the Borrowdale volcanic rocks are caught up in the igneous rock. The Skiddaw Slates are somewhat altered, and in some cases extremely fine veins—often several in the thickness of an inch—are forced along the divisional planes of the slate producing a schistose rock by *lit-par-lit* injection.

Some of the fragments of the Borrowdale rocks contain garnets of the usual type, and similar garnets are occasionally found in the microgranite, whether as original crystals or by derivation from the volcanic rocks is not clear.

Radial aggregates of tourmaline are found on some of the joint-faces of the microgranite. This mineral has not been extensively noted in the district, though recently recorded by T. A. Jones in the Eskdale granite (see *Geol. Mag.* Decade VI, vol. II, p. 190).

Harker describes the Armboth and Helvellyn dykes as follows: they are "garnetiferous in places, are microspherulitic rocks, with

[1] Harker, *loc. cit.*

porphyritic crystals of quartz and felspar, which serve as nuclei for spherulitic growths."

The Blea Crag rock will be referred to again presently.

The granophyre of Ennerdale and Buttermere is according to Harker "a pink fine-textured rock with indistinct quartz-grains and crystals of felspar. Thin slices shew it to consist mainly of micropegmatite. There are chlorite pseudomorphs after augite and biotite." A specimen from Scale Force has a silica percentage of 71·442. The age of this rock is doubtful, but from study of its effects upon the earth-movements it appears to have been injected before the great movements at the end of Lower Palaeozoic times, and it is probably connected with igneous activity of the period of the Borrowdale volcanic rocks. It is of the nature of a complex laccolith[1].

Two at least of the granophyric masses give evidence of being products of magmatic differentiation, namely the Blea Crag rock in the Langstrath Valley and the Ennerdale-Buttermere rock. Evidence of this in both cases is furnished in E. E. Walker's paper to which reference has already been made, and in the case of the latter rock further information is supplied in Rastall's paper.

The Blea Crag granophyre shews a micropegmatitic intergrowth of quartz and felspar. It frequently contains garnets. A typical example gives a silica percentage of 64·4. Associated with it is a black, more basic rock with 57·91 of silica. There is evidence that these rocks were the result of magmatic differentiation followed by intrusion firstly of the more basic rock, then of the granophyre, and that the interval between the two periods was short. By an intermixture of the extreme types intermediate types were produced, containing felspar pheno-crysts, and having a composition intermediate between those of the extreme types. One of these intermediate rocks has a silica percentage of 60·02, and another of 61·63.

Walker also described a series of more basic rocks associated with the Ennerdale granophyre, and these have been more fully considered by Rastall. They occur in Burtness Combe, south-west of Buttermere. There is evidence that these also were intruded before the granophyre, and that rocks of mixed type resulted from the action of the acid rock upon the basic. Rastall says "the earliest phase of intrusion is that of small masses of more basic character, which now occupy a marginal position; these may be described as ranging from dolerites and quartz-dolerites to a type in which the presence of a considerable proportion of orthoclase shews affinities to the quartz-monzonites of Prof. Brögger's classification. This was quickly followed by the intrusion of the main

---

[1] A full account of the Ennerdale-Buttermere series of intrusive rocks has been given by R. H. Rastall, *Quarterly Journal of the Geological Society*, 1906, vol. LXII, p. 253.

mass, while the earlier injections were in some cases still hot and partly liquid." It will be seen that there is much similarity in the sequence of events in the case of the Blea Crag and Buttermere rock-complexes. Rastall also notices a modification of the granophyre at a place one mile west of Scale Force, where a greisen is found; he also mentions that the granophyre is traversed by aplite veins on the south side of Ennerdale.

One other point in connexion with the intrusive rocks of the older group may be noticed. Harker[1] gives a list of a series of sills and laccoliths injected into the Ordovician rocks, which are so arranged that those lower in the series are denser than those which are at a higher level. He notices the differentiating action of gravity in effecting a stratiform "arrangement of layers of different density in a magma still wholly fluid" and suggests that this may have an application to these sills and laccoliths. The two highest sills which he mentions are in the Coniston Limestone Series, but this series also contains contemporaneous volcanic rocks, and the usual interpretation of the sequence in the Lake District allows that these are the final effects of the phase of volcanic activity which attained its maximum during the formation of the rocks of the Borrowdale Series.

In connexion with the intrusive igneous rocks mention may be made of various metalliferous and other veins. It is now recognised that many ore deposits are genetically connected with intrusive rocks, some being actually formed during the consolidation of the rock, while others have been derived from the rocks owing to the liberation of vapours, including steam, after consolidation while the rock is still at a high temperature[2].

It is probable that some of these veins in the district were formed in connexion with the intrusive rocks of Ordovician date, though others were clearly formed in later times.

Among the rocks of the Skiddaw Slates and the Borrowdale Series we find numerous metalliferous lodes, supplying ores of copper and lead, and in much smaller quantity of zinc, cobalt and antimony. The chief mines in the Skiddaw Slates have been worked in and around the Vale of Newlands, and on the south side of Saddleback, while the most important of those

---

[1] Harker, A., *Quarterly Journal of the Geological Society*, 1895, vol. L, pp. 131, 132 and footnote on latter page.

[2] A paper by J. D. Kendall on "The Mineral Veins of the Lake District" appeared in the *Transactions of the Manchester Geological Society*, 1884, vol. XVII, pp. 292–341. A separate work on the subject "Mines and Mining in the Lake District" was written by J. Postlethwaite.

of the Borrowdale Series are at and near Coniston, and on the east side of Helvellyn. The celebrated graphite mine of Borrowdale is described by Clifton Ward in the Memoir on "The Geology of the Northern Part of the English Lake District," where will also be found an account of the metalliferous mines of that part of the district. The graphite "occurs in close connection with a...highly altered diorite lying between two other masses of intrusive blue trap (diabase) of a compact character."

In the case of the mineral veins even more than in that of the intrusive igneous rocks it is difficult to assign exact dates to their formation.

The relative abundance of metalliferous veins in the Skiddaw Slates and Borrowdale Series, as opposed to their extreme rarity and very local distribution in the Silurian rocks, suggests the pre-Silurian date of many of them. Furthermore, the fact already noticed that many of these veins were filled by vapours issuing from igneous rocks suggests that those of the Lake District which are in proximity to the igneous rocks described in this chapter have a genetic connexion with them.

At Shap there is an interesting case of a metalliferous vein containing galena and pyrites which has been metamorphosed by the Shap granite, and was therefore formed earlier than the intrusion of that rock. It is therefore of a date prior to the particular period of Devonian times when the Shap granite was consolidated, and is probably of Lower Palaeozoic age.

Much speculation has arisen concerning the origin of the graphite of the Seathwaite mine. As suggested by Ward it is most probably connected with the diorite (?) with which it is associated, for he states that it actually occurs in that rock "in black specks or forming a light blue-black, nebulous haze" as well as in the veins bordering the rock. Whence the carbon was derived has not been ascertained. It has been suggested that it came from material furnished by graptolites in the Skiddaw Slates, but no definite evidence has been advanced in favour of this view.

It must not be supposed that the effects of vapour and heated waters were confined to the rocks in mineral veins. Many of the volcanic rocks of the Borrowdale Series have undergone changes due to hydrothermal action with the production of such minerals as quartz, chlorite and epidote. These minerals often vein the rocks and produce alteration in the adjoining rock, which is lighter in colour owing to the infiltration of quartz and epidote. Walker describes veins of epidosite (a quartz-epidote rock) in the intrusive rocks of the Langstrath Valley, and they are abundant in many of the contemporaneous volcanic rocks.

# CHAPTER V

## LOWER PALAEOZOIC ROCKS

### C. *The Coniston Limestone.*

The general trend of these rocks from Shap to Millom has been noticed in the introductory chapter. Along this line, which is often broken by transverse faults, the beds strike in an east-north-east and west-south-west direction from Shap to the head of Coniston Lake; thence to Millom the strike changes to north-east and south-west. The Coniston Limestone re-appears on the east side of the Duddon Valley between Ireleth and Dalton-in-Furness. Lastly a remarkable development of rocks of this age differing markedly in their characters from those of the main line of outcrop occurs far away to the north among the Caldbeck Fells, at the head of the little valley of Dry Gill.

As was the case with the junction between the rocks of the Borrowdale Series and the Skiddaw Slates, so here we find a difficulty in understanding the nature of the junction between the Borrowdale volcanic rocks and the Coniston Limestone. It has long been known that the strike of the two sets of rocks is discordant, but this discordance might be due either to an unconformity or to faulting. The discordance of strike is noticeable in places along the whole line of junction but it becomes particularly pronounced in the country to the west of Coniston Lake.

This is clearly brought out on the map (facing p. 17) shewing the divisions of the rocks of the Borrowdale Series. It will be noticed that in passing south-westward from Coniston, members of the Shap rhyolites, Shap andesites, Scawfell Ash and Breccia group and Ullswater group abut against the base of the Coniston Limestone. The discordance along this line was ascertained as the result of the detailed mapping of the geological surveyors. The explanatory memoir of the survey map of this portion of the district has not been published, but the results obtained are summarised by the late Mr De Rance in the discussion of a paper read before the Geological Society in 1877[1]. He remarked "that in tracing the outcrop of the Coniston Limestone across country, it was found to rest upon different members of the underlying volcanic series, which plunge under it with varying direction of strike and amount of dip, the unconformity being so marked between the two sets of rocks that occasionally the volcanic series appear to have obtained a dip, been denuded, and faulted before the deposition of the overlying Coniston series." This view of unconformity has been advocated by other members of the Geological Survey, and more recently by J. F. N. Green. On the other hand Dr Harker and myself when engaged in work on the district obtained evidence which led us to believe that we were here dealing with a fault, which, like that between the Skiddaw Slates and Borrowdale Series, makes a low angle with the horizontal plane[2], but of this more anon.

In the table on p. 7 the Coniston Limestone is stated to have a thickness of from two hundred to a thousand feet. The latter thickness occurs where some contemporaneous lavas are associated with the sediments between Shap Wells and Kentmere; elsewhere the thickness of the division approximates more nearly to the lower estimate.

Though the name Coniston Limestone is given to the rocks of this division, limestones actually play a subsidiary part in its formation, and the limestone bands which do occur are usually impure; but the name is convenient as marking the abundance of calcareous matter (often forming definite limestone bands) in this division, as contrasted with its scarcity in the rest of the Lower Palaeozoic rocks of the district. The rocks consist essentially of calcareous mudstones often containing a considerable proportion of fine volcanic dust, which becomes coarser where the contemporaneous lavas are developed. In these mudstones limestone bands of varying degrees of purity occur, but they are often found as a series of discontinuous nodules running in bands along the planes of stratification. The highest member of the series is a very constant blue-grey mudstone, known as the Ashgill shale.

The contemporaneous lavas between Shap Wells and Kentmere are rhyolites, often spherulitic and generally of a pink colour. They frequently exhibit good examples of flow

---

[1] See *Quarterly Journal of the Geological Society* (1877), vol. XXXIII, p. 483.
[2] Marr, *Proceedings of the Geologists' Association*, 1900, vol. XVI, p. 449.

structure. Their resemblance to the rhyolites of the Shap rhyolite group at the top of the Borrowdale Series is noteworthy.

Study of the fossil contents of the Coniston Limestone shews that it is divisible into two main groups, a lower which I have termed the Sleddale group and an upper, the Ashgill group[1].

The Sleddale group is of the age of the Caradoc rocks of North Wales. The Ashgill group represents a series which is typically developed in the north of England, and the name *Ashgillian Series* has been given to it.

The fossils of the Coniston Limestone often occur in extraordinary abundance. We meet with corals, brachiopods and trilobites in profusion and more occasionally representatives of other animal groups. Large brachiopods of the genera *Orthis*, *Strophomena* and *Leptaena* are frequent, and among the trilobites the genera *Trinucleus, Calymene, Cybele* and *Phacops* predominate. The species found in the lower and upper members of the Coniston Limestone are those which respectively characterise the Caradocian and Ashgillian divisions of Wales and other districts.

The following is a classification of the rocks of the Coniston Limestone group:

|  |  | feet |
|---|---|---|
| | Ashgill shales ... ... ... ... | 50 |
| | *Phacops-mucronatus* beds ... ... | 16 |
| Ashgillian Group | Volcanic ash ... ... ... ... | 16 |
| | White limestone ... ... ... | 12 |
| | *Phillipsinella* beds ... ... ... | 7 |
| | Applethwaite beds with ... ... | to 100 |
| Sleddale Group | Conglomerate at base ... ... ... | 10 |
| | Contemporaneous rhyolites ... ... | variable |
| | Stile End beds ... ... ... ... | about 50 |
| Roman Fell Group | Not represented in the district? | |

Beginning with the Stile End beds we may first consider the nature of the junction between them and the underlying Borrowdale rocks.

[1] Marr, "The Coniston Limestone Series," *Geological Magazine*, Decade III, vol. IX, p. 97. (In the adjoining area east of the Eden Valley a still lower group, the "Roman-Fell Group," is developed. It has not been detected in the Lake District.)

An excellent exposure of this is seen at High Pike Haw, south-west of Ashgill Quarry, Coniston. The following section (Fig. 7) shews the complete succession at this place from the top of the Borrowdale Series to the top of the Ashgill shales.

The highest rocks referred to the Borrowdale Series are vesicular lavas succeeded by banded ashes and agglomerates dipping in a south-westerly direction, that is at right angles to the overlying rocks. The remaining rocks dip to the south-east. The lowest rock seen with this dip is a green ashy looking rock, which is succeeded by a purple breccia[1]. These may mark an unconformity or a fault. Above the breccia is a series of ashy grits and conglomerates overlain by calcareous ashy rocks of the characters of the Stile End beds of Stile End, and containing fossils. A conglomerate rests upon this and apparently forms the base of the upper part of the Caradocian succession. Above this the

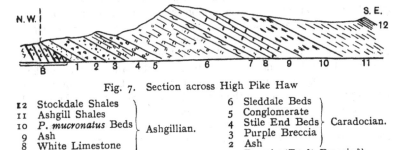

Fig. 7. Section across High Pike Haw

| | | | | |
|---|---|---|---|---|
| 12 | Stockdale Shales | | 6 | Sleddale Beds |
| 11 | Ashgill Shales | | 5 | Conglomerate |
| 10 | *P. mucronatus* Beds | Ashgillian. | 4 | Stile End Beds |
| 9 | Ash | | 3 | Purple Breccia |
| 8 | White Limestone | | 2 | Ash |
| 7 | *Phillipsinella* Beds | | 1 | Breccia (Fault Breccia?). |

6, 5, 4, 3, 2, 1 — Caradocian.

B Borrowdale Series.

Applethwaite beds of Caradocian age and the various representatives of the Ashgillian series follow in normal sequence.

J. F. N. Green, in the paper to which reference has been made, describes junctions of the Coniston Limestone with the rocks of the underlying Borrowdale Series in the part of country around Ireleth, and as already remarked, he considers that the appearances there shew that the former rocks lie unconformably upon the latter.

The reference of the lowest rocks of the Coniston Limestone of the south-western portion of the district to the Stile End group is purely inferential, being based on lithological characters, and the occurrence of a conglomerate at the base of the Applethwaite Limestone division. We may turn to the eastern part of the district where the typical development of Stile End rocks is found, to note their general characters. It has already been seen that contemporaneous volcanic

---

[1] I find that the green rock is recorded in my note-book as an ash. This note was made some twenty years since, and I have not re-examined the rock. [Purple breccia occurs below No. 2, 1915.]

rocks are found between Shap Wells and Kentmere. Below these is a set of fossiliferous sediments to which Professors Harkness and Nicholson gave the name of Stile End [Grassing] beds, from farms which lie on the uplands between Kentmere and Long Sleddale. These sediments consist of grey calcareous ashes of fine grain weathering to a rusty brown colour. They contain badly preserved fossils, chiefly corals, large brachiopods and specimens of *Phacops* of the subgenus *Chasmops*. The fauna though poor seems identical with that of the Applethwaite beds. A similar scanty fauna is found in the beds of the south-west of the district which we have referred to the Stile End division.

The volcanic rocks above these Stile End beds between Shap Wells and Kentmere are, as already stated, of rhyolitic character, and shew marked resemblance to the rhyolitic rocks of the uppermost division of the Borrowdale Series. On Yarlside pyroclastic rocks are developed including an extremely coarse agglomerate, containing fragments of rhyolite, some of a nodular character, others converted into jasper. The rhyolitic flows are also well-developed on Yarlside, and are further exhibited in the stream of Stockdale which flows into the Long Sleddale stream. These lavas have a flow structure, the lamellar lines of flow being often highly contorted as the result of differential movement prior to the consolidation of the rock. Much of the rock is spherulitic[1], and often nodular. The nodules are usually about an inch in diameter but in places attain a size of several inches. These nodules have sometimes been replaced by quartz, and the replacement occasionally extends through the rock around the nodules. The rhyolite is in places converted into jasper.

The Applethwaite beds immediately succeed the rhyolitic series above described. The junction is well shewn in Stockdale Beck and its tributary Browgill. The highest member of the rhyolitic group is there an ash. Upon it rests a detrital rock consisting of a calcareous ashy matrix with subangular fragments chiefly composed of rhyolite. In Browgill the highest bed of this conglomeratic division has a more purely calcareous matrix containing mostly well-rounded pebbles of rhyolite. This passes insensibly into the overlying non-conglomeratic sediments of the Applethwaite group.

As already stated a similar conglomerate occurs at High Pike Haw, and has also been found in many other exposures in the central and south-western parts of the district.

The Applethwaite beds like those of Stile End contain much ashy sediment of a greyish hue weathering brown. The amount of calcareous

---

[1] A description of these rocks with admirable illustrations of their microscopic structure by F. Rutley will be found in the *Quarterly Journal of the Geological Society* (1884), vol. XL, p. 340.

sediment in this division is greater than among the Stile End beds, and there is no doubt a considerable amount of mechanical sediment which is only indirectly of volcanic origin, having been derived from earlier rocks by erosion.

The fauna of these beds is very rich. Corals of the genera *Lindstroemia* and more rarely *Heliolites* and *Syringophyllum* are found. Large brachiopods of genera already named, with others, are very abundant. *Chasmops*, *Cybele*, and *Calymene* are most abundant among the trilobites though *Trinucleus* is not rare in places. No graptolites have yet been found in this division in the district, and they should be sought out. Many fossil lists of these beds are recorded in Survey Memoirs and elsewhere. These lists in all cases contain the names of some fossils which belong to the overlying Ashgillian Series. Now that the rocks of the latter series have been separated from those of Caradocian age, it is necessary that some geologist should undertake the revision of the lists. This will require much work in the field as well as in the museum.

I have recently studied the division of the Ashgillian strata at Ashgill and the tract of country in its neighbourhood. The subdivisions which I have established are given in the table on p. 36.

The lowest beds (the *Phillipsinella* beds) are lithologically similar in general appearance to those of the underlying Applethwaite division of the Caradocian strata, and contain much fine ashy matter. They are blue instead of grey, and are usually more closely jointed. The fauna is however very different, though some species come up from Caradocian strata. The genus *Phillipsinella* is very abundant. *Staurocephalus* is far from rare, and other species occur which are found in Ashgillian strata of various districts. The brachiopods, unlike those of the underlying strata, are usually of small size.

The succeeding White Limestone is a very prominent band in the field, usually standing out conspicuously. It has hitherto yielded few fossils, which are undoubtedly of Ashgillian age, and recall those of the well-known Keisley Limestone of the east side of the Eden Valley. There are also certain lithological resemblances between the two rocks. I suspect that the massive crystalline White Limestone of the Millom district is an expanded development of this band.

The ash which succeeds the limestone is a fine-grained rock. It is of importance as a readily recognisable band by means of which we can establish our position in the sequence. It is only developed, so far as is known, in the south-western part of the district, not having been detected to the east of Coniston Lake. It is also of importance as occurring on about the same horizon as a more fully developed series of contemporaneous volcanic rocks in the neighbourhood of Sedbergh, shewing that in the Lake District, as near Sedbergh, contemporaneous volcanic action continued until near the close of Ordovician times.

The *Mucronatus* beds are greenish calcareous mudstones, often formed very largely of the species of *Phacops* from which the beds have been named: the genus *Staurocephalus* is also found in these beds, and many other fossils of Ashgillian age occur in them.

There is a gradual passage, by loss of calcareous matter, from these beds into the Ashgill shales, the highest beds of the Ordovician succession.

These beds are leaden-blue mudstones, usually breaking along divisional planes into prismatic fragments, though at Ashgill itself they are so highly cleaved that they have been worked as roofing slates. The base of the shales often contains a considerable proportion of fine gritty material. The shales contain numerous fossils, chiefly brachiopods, of which *Strophomena siluriana* is characteristic. *Phacops mucronatus* is also found though not abundantly. In parts of Appletreeworth Beck the topmost bed contains abundant ostracods, so that it resembles a pisolite.

Fragments of graptolites have been found in the Lake District in beds which are almost certainly Ashgillian, though at the time of their discovery these beds were not separated from the underlying Caradocian strata. I suspect them to occur in the White Limestone and possibly also in the *Mucronatus* beds, but all divisions of the Ashgillian strata should be carefully examined in hopes of discovering them. The typical Ashgillian graptolite found in the Cautley district near Sedbergh is *Dicellograptus anceps*. One feels confident that this form will also eventually be found in the Ashgillian rocks of Lakeland.

Though the various divisions of the Ashgill shales have hitherto only been studied in detail in the tract to the south-west of Coniston village, it is probable that the same succession will be found all along the outcrop of the Coniston Limestone. I have found probable *Phillipsinella* beds, the White Limestone, the lower part of the *Mucronatus* beds with Cystidea, and the upper part of the same beds with abundant *P. mucronatus* at Skelgill, succeeded by the Ashgill shales. The beds with Cystidea were exposed in a section between Skelgill and Troutbeck which is now covered over. Specimens of these Cystidea are also preserved in the Kendal museum from the east side of Troutbeck. The ash between the White Limestone and the *Mucronatus* beds has not been found hitherto to the east of Coniston Lake, and I believe that it is not represented there.

The gritty beds with brachiopods which are found near the top of the Ashgill shales in the Sedbergh district have not been detected in Lakeland. There are fossils in the Kendal museum from Rebecca Hill near Ireleth, in a rock which resembles that of the Sedbergh area. *Orthis protensa* and *Strophomena siluriana* are among these fossils.

The higher Ordovician and lower Silurian strata of this tract on the east side of the Duddon estuary require further study. In the

Sedgwick museum is a specimen of a pale green rock with *Stricklandinia lirata*. It was found at Rebecca Hill. Its horizon is not known, and it may be an Ashgillian fossil, though the character of the rock suggests that it came from Browgill beds of Valentian age.

The Ashgillian faunas of Lakeland require much further study.

I have left for separate consideration the interesting development of beds of Caradocian age in Drygill, a stream west-north-west of the summit of Carrock Fell[1]. The position of the beds is shewn in the

Fig. 8. Sketch map of Drygill.

*d* Drygill Beds.      *e* Eycott Volcanic Rocks.      *S* Skiddaw Slates.
*F, D, G, g* Intrusive Igneous Rocks.

accompanying sketch map (Fig. 8). South of the intrusive rocks of Carrock Fell are the Skiddaw slates, while to the north of the Drygill shales, the volcanic beds of the Eycott group dip away to the northward. The Drygill shales themselves are dipping to the south.

They consist of grey-black shales weathering to a dirty-white hue. They are fine grained, and contain when unweathered a fair amount of

[1] These beds were described by Professor Nicholson and the present writer in the *Geological Magazine* (1887), Decade III, vol. IV, p. 339.

calcareous matter. Some igneous rocks of acid composition (quartz felsites) are probably intrusive but may be contemporaneous. The shales are also penetrated by mineral veins containing the rare mineral campylite in small barrel-shaped crystals. Fossils are extremely abundant, especially trilobites and brachiopods. Among the fossils are *Ampyx rostratus*, *Trinucleus concentricus* and *Orthis testudinaria*.

The position of these beds induced Prof. Nicholson and myself to refer them to an earlier age than their true one, but subsequent consideration of the fauna led me to infer that they were of newer age. This was amply confirmed later by Miss Elles and Miss Wood[1].

These authors, from a further study of the organic remains, of which a list is given in their paper, shewed that the strata were Caradocian; and the occurrence of certain forms as *Ampyx tumidus* and *Phacops appendiculatus* (= *mucronatus*) suggests the possibility that equivalents of Ashgillian strata may also be present. Lithologically the beds are unlike those developed in the south of the district. They strongly resemble a group of rocks chiefly of Caradocian age (in which however representatives of the Ashgillian series may also be present), which are developed on the east side of the Eden Valley, where they are known as the Dufton shales. The main point of interest of these beds is their geographical position, which bears upon the question of the nature of subsequent earth-movements. This will be discussed in a later chapter.

# CHAPTER VI

## LOWER PALAEOZOIC ROCKS

### SILURIAN SYSTEM

#### D.   *The Stockdale Shales.*

It seems to be established beyond doubt that, in the Lake District, the Silurian rocks succeed those of Ordovician age with perfect conformity. The sequence is frequently exposed, as at Browgill, Skelgill near Ambleside, Ashgill Quarry and Appletreeworth Beck, south-west of Coniston. In each case the beds above and below the junction have the same dip and strike. Where unfaulted, the lowest member of the Stockdale

---

[1] Elles, G. L. and Wood, E. M. R., "Supplementary Notes on the Drygill Shales," *Geological Magazine* (1895), Decade IV, vol. II, p. 246.

shales always rests upon the Ashgill shales, and apparently also upon their highest bed.

Nevertheless there is no sign of passage; the lithological change is abrupt and complete. There is also a marked contrast between the fauna of the Ashgill shales and that of the Stockdale shales. At least one species, *Phacops mucronatus*, is common to the two, but the majority of forms are distinct. It must be remarked, however, that the fauna of the Stockdale shales is mainly one of graptolites, and that these creatures have not been found in the Ashgill shales. From what we know of the change in the faunas at this horizon in other districts, as in the southern uplands of Scotland, we may infer with confidence that, in our district also, there was a sudden change in the general nature of the faunas at the plane of junction separating the Ordovician from the Silurian strata.

In Lakeland the Stockdale shales are comparatively thin, having a maximum thickness of about 250 feet. There is evidence that they were accumulated very slowly, and it is known that they represent deposits of much greater thickness elsewhere, namely the Llandovery and Tarannon groups of the Welsh Borderland.

They are divisible into two groups, the lower, known as the Skelgill beds, being typically developed in a stream near High Skelgill Farm, near Ambleside, and the upper group termed the Browgill beds, from a gill whose stream is a tributary of Stockdale on the east side of Long Sleddale: it is from Stockdale that the whole series takes its name.

The upper division is much thicker than the lower one. The latter has a thickness of about 50 feet, while the former is about four times as great, but there is evidence that the Skelgill beds were deposited much more slowly than those of Browgill age, and the thinner Skelgill beds probably represent a longer period of time than do the thicker Browgill deposits.

The Skelgill beds consist of black graptolite-bearing shales containing thin light green laminae of subordinate thickness, interstratified with pale blue, slightly calcareous mudstones marked by absence of graptolites, but containing remains of organisms of other groups which are usually absent from the graptolite-bearing shales.

One of these non-graptolitic deposits forms the base of the division, in the western part of the area, and is much more calcareous than those at a higher horizon.

The Browgill beds are largely composed of greenish shales, resembling those which occur in a minor degree as laminae in the Skelgill beds. Interstratified with these are graptolite-bearing shales, usually of somewhat lighter colour than those of the Skelgill beds. These shales play a very subordinate part in the Browgill beds. In addition there are fairly thick masses of greenish grit.

The higher part of the Browgill beds is usually stained reddish-purple. It is not quite clear whether this colouring matter was introduced when the deposits were formed, or as the result of subsequent change, though one suspects the latter.

The graptolites of the Skelgill beds form a very different assemblage from those which are found in the Ordovician strata. Many of the Ordovician genera have disappeared, though some, as *Climacograptus* and *Diplograptus*, are still fairly abundant. The particularly characteristic Silurian genus *Monograptus* appears near the base of the deposits; other Silurian genera are *Rastrites*, which is confined to Llandovery-Tarannon strata, and *Dimorphograptus*, which is limited to the lower part of the Skelgill beds and their equivalents elsewhere.

The genera above-mentioned, except *Dimorphograptus*, also occur in the Browgill beds, and another genus *Cyrtograptus* here appears, and passes upward into higher strata.

The fossils other than graptolites are small corals, brachiopods, trilobites, phyllocaridan crustacea and cephalopods. The trilobites are of the genera *Cheirurus, Phacops, Encrinurus, Harpes, Ampyx* and others. The phyllocaridan crustacea and cephalopods are found in the graptolite-bearing shales, the others chiefly in the interstratified blue mudstones.

The graptolites occur in fossil zones, each zone being characterised by a definite species, and also by a general assemblage of forms differing in some respects from those of overlying and underlying zones. Seven such zones have been detected in the Skelgill beds and only two in the Browgill beds, but the highest zone of the Skelgill beds is absent from the Lake District, though it is developed in the Sedbergh region.

The principal fossiliferous localities for the Skelgill beds are Browgill, Skelgill, Church Beck and its tributary Mealy Gill, Coniston, and Appletreeworth Beck. Those of the Browgill beds are Browgill, Pull Beck west of the head of Windermere, and Ashgill Beck.

There is little doubt that the Skelgill beds were deposited in water of considerable depth. This is shewn by their small thickness, and the fact that they are represented by normal sediments of much greater thickness elsewhere; by the great changes which take place in the faunas in this limited thickness, as indicated by the different assemblages of organisms of the different fossil zones, and possibly by the dwarf character of most of the organisms other than graptolites. The characters of the strata point to the deposit of fine mud at a considerable distance from the coastal margins. Professor Lapworth has given reasons to suppose that the graptolites were organisms living in the surface-waters, attached to floating weeds of a "Sargasso" sea, and that dead fragments of weed and graptolites were showered down on to the sea-floor. He supposes that the carbonaceous matter which causes the blackness of the mudstones was derived from these weeds. The alternation of black graptolite-bearing shales with blue non-graptolitic mudstones would indicate the alternation, owing to climatic or other changes, of periods when the "weed" grew abundantly and the black shales were laid down, and other periods when the "weed" was scarce or absent and the blue mudstones were accumulated.

Somewhat similar conditions must have prevailed during the formation of the Browgill beds, but the waters were now shallower and nearer the coast, as shewn by the inclusion of gritty deposits, and the periods of "weed" growth were less numerous and shorter than during the formation of the Skelgill beds, as proved by the fewness and thinness of the graptolitic bands in the former beds.

Some further details concerning the nature of the various divisions of the Stockdale shales may be given. The following is a classification of their beds[1]:

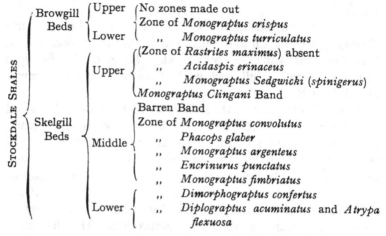

Fig. 9 gives a section across the typical development of the Skelgill beds of Skelgill and Fig. 10 that across the type of the Browgill beds at Browgill.

Beginning at the base, the Ashgill shales are usually succeeded by a mottled grey limestone from four inches to a foot thick, with brachiopods, including the type-species *Atrypa flexuosa*, but at Browgill the basal deposit is more argillaceous and contains graptolites, including *Diplograptus acuminatus* which elsewhere characterises the basal zone of the equivalent strata.

These are succeeded by very black mudstones from 12 to 15 feet thick, with *Dimorphograptus confertus* and other species, *Monograptus revolutus*, and several other graptolites.

A strike fault often occurs, either in the *Dimorphograptus* beds, or between them and higher deposits; accordingly an exact estimate of the thickness is impossible. In the tract south-west of Coniston the beds of this zone appear to have a much greater thickness: they are however much contorted, and the apparent greater thickness is at any rate largely due to subsequent repetition by folding.

The lowest member of the middle Skelgill division also consists of very black graptolite-mudstones characterised by *Monograptus fimbriatus*. This and the two graptolite-bearing shales at a higher horizon also

---

[1] They were described in a paper by the late Professor Nicholson and the present writer in the *Quarterly Journal of the Geological Society* (1888), vol. XLIV, p. 654.

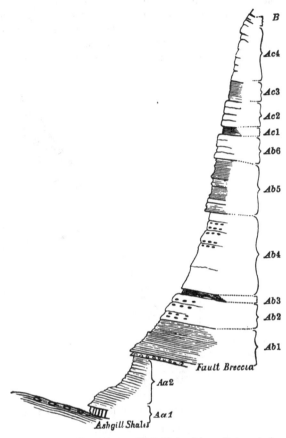

Fig. 9. Section across Skelgill (scale 15 ft to 1 in.).

| | |
|---|---|
| *B* | Browgill Beds. |

| | | |
|---|---|---|
| *Ac*4 | Zone of *Acidaspis erinaceus* | ⎫ |
| *Ac*3 | ,, *Monograptus Sedgwicki* | ⎬ Upper Skelgill Beds. |
| *Ac*2 | ,, *Ampyx aloniensis* | |
| *Ac*1 | *Monograptus Clingani* Band | ⎭ |

| | | |
|---|---|---|
| *Ab*6 | Barren Band | ⎫ |
| *Ab*5 | Zone of *Monograptus convolutus* | |
| *Ab*4 | ,, *Phacops glaber* | ⎬ Middle Skelgill Beds. |
| *Ab*3 | ,, *Monograptus argenteus* | |
| *Ab*2 | ,, *Encrinurus punctatus* | |
| *Ab*1 | ,, *Monograptus fimbriatus* | ⎭ |

| | | |
|---|---|---|
| *Aa*2 | ,, *Dimorphograptus confertus* | ⎫ Lower Skelgill Beds. |
| *Aa*1 | ,, *Atrypa flexuosa* | ⎭ |

contain *Monograptus gregarius*, so that the three shale-bands, with their intervening mudstones, represent Prof. Lapworth's *gregarius*-zone in the typical development of the Birkhill shales of the southern uplands. The beds probably have a thickness of about ten feet. They are succeeded by non-graptolitic mudstones with *Encrinurus* which

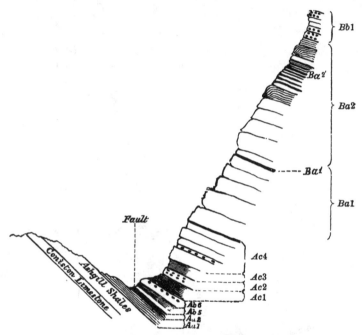

Fig. 10. Section across Browgill (scale 24 ft to 1 in.).

Bb1  Lower part of Upper Browgill Beds.
Ba2  Zone of *Monograptus crispus* (fossil bands at *Ba²'*).
Ba1   ,,   *Monograptus turriculatus* (fossil band at *Ba¹'*).

Ac4   ,,   *Acidaspis erinaceus*
Ac3   ,,   *Monograptus Sedgwicki*
Ac2   ,,   *Ampyx aloniensis*
Ac1  *Monograptus Clingani* Band
Ab6  Barren Band                              } Skelgill Beds.
Ab5  Zone of *Monograptus convolutus*
Aa2   ,,   *Dimorphograptus confertus*
Aa1   ,,   *Diplograptus acuminatus*

require no further notice. The succeeding bed, the *Monograptus argenteus* zone, is of great interest. It is only eight inches thick, and again consists of very black shales. In addition to the type fossil and *Monograptus gregarius*, we have a swarm of other species, for this is by far the richest zone in the succession, and moreover, the fossils are usually excellently preserved in a state of high relief. Perhaps the most interesting feature of this zone is the occurrence of a band of green non-graptolitic shale near the centre: I have frequently alluded to this as the "green streak." It is seen at Skelgill, Mealy Gill and Appletreeworth Beck, and it is also found in the Sedbergh district, always being in the same position and having approximately the same thickness. Prof. O. T. Jones has detected what is almost certainly the same band in the country east of Aberystwyth, though it is there about an inch thick, as we should expect, since the whole of the representatives of the Stockdale shales are there thicker than in the Lake District. The wide distribution of this streak is another argument in favour of these deposits having been laid down on an ocean-floor at some distance from the coast-margins.

The succeeding mudstones with *Phacops glaber* and other trilobites are about twelve feet thick. They are marked by a large number of calcareous nodules which usually weather out in the cliffs. Others of these beds contain similar nodules, but in fewer bands than in the case of the one under consideration.

These mudstones are overlain by another graptolitic zone, that of *Monograptus convolutus*. It is about eight feet thick, and marks a change in the characters of the graptolitic sediments. Those below are black, while this and the succeeding one are generally of a prevalent blue-grey colour, with black bands of subordinate thickness in places. A band of blue mudstones above these, four feet thick, and containing few fossils, is spoken of as the Barren Band. It is taken as the uppermost deposit of the Middle Skelgill beds.

The Upper Skelgill beds have at the base a graptolitic band about a foot thick. It contains, so far as known, no characteristic species, but, from the abundance of *Monograptus Clingani*, is known as the *Clingani* band.

Four feet of non-graptolitic mudstones succeed. They are character-ised by *Ampyx aloniensis*. This form is of interest. The genus is typically Ordovician, and occurs but rarely in the Silurian, though it is known in beds of Ludlow age near Malvern.

A series of vari-coloured graptolite shales succeeds. These shales contain swarms of the typical *Monograptus Sedgwicki* (= *spinigerus*) with many other graptolites.

The highest division of the Skelgill beds in Lakeland (where as we have seen the zone of *Rastrites maximus* is unrepresented) consists of ten feet of blue non-graptolitic mudstones, characterised by the very beautiful *Acidaspis erinaceus*, which has been found most abundantly in Torver Beck, Coniston.

It may be remarked that the principal localities for fossils other than graptolites are Browgill, Skelgill, Pull Beck and Torver Beck.

Turning now to the Browgill beds, we find about twenty feet of pale green shale with a band at the summit marked by *Monograptus turriculatus*. These are all referred to the zone of that name, and as in the Sedbergh district several bands containing these fossils have been found, they will probably be detected in Lakeland also. They are succeeded by about forty-four feet of green shales with a number of graptolitic bands most of which occur within a thickness of fifteen feet, having nineteen feet of graptolitic shales below and ten feet above. These graptolite shales are characterised by *Monograptus crispus* and *Monograptus exiguus*, and the remarkable *Retiolites Geinitzianus*, which passes up into Wenlock beds, is frequent in them. The whole forty-four feet is referred to the *M. crispus* zone.

The upper Browgill beds consist of seven or eight feet of blue calcareous mudstones succeeded by about sixty feet of pale green and reddish purple shales, with massive grit beds of a grey-green colour. As already stated, no fossils have yet been detected in these beds.

# CHAPTER VII

## LOWER PALAEOZOIC ROCKS

### SILURIAN SYSTEM

E.   *The Lower Coniston Flags = Brathay Flags.*

It will be seen, by consulting the table on p. 7, that the group of rocks to which the name Coniston Flags has been given are divisible into lower and upper divisions, of which the former represent the Wenlock beds of the Welsh Borders, while the latter are equivalent to the lower portion of the Lower Ludlow rocks of the same region. The Lower and Upper Coniston Flags were naturally grouped together in the days when the geology of the district was first studied, having the very obvious feature in common, that both consist largely of flaggy beds. Even from a lithological point of view, however, there are many features wherein the Upper Coniston Flags differ from the lower strata of the same group, and resemble the strata of the succeeding Coniston Grits. The separation of lower from upper flags, and the union of the

latter with the Coniston Grits appear natural when we study
the organic contents of the strata. It will be convenient
therefore to consider the Lower Coniston Flags in a separate
chapter, and to deal with the equivalents of the Ludlow beds
in a later one. The name Brathay Flags was applied to these
Lower Coniston Flags by Sedgwick; it is a term which may
well be adopted, for we certainly require a name whereby we
may distinguish the local deposits of Wenlock age from those
appertaining to the Ludlow Series.

The Brathay Flags crop out along a strip of country of no
great width lying south and south-east of the line of outcrop
of the Stockdale shales, between Shap Wells and Millom, and
they reappear to the east of the outcrop of the same shales on
the eastern side of the Duddon estuary. The average thickness
of the beds is about 1000 feet, but the thickness varies
considerably, not only because of original variations but also
on account of differences produced by the varying effects of
subsequent lateral pressure.

The beds originally consisted of blue-grey laminated mudstones, sometimes containing thin partings of fine grit. As a
result of subsequent pressure they have been converted into
slates, which have in places been extensively worked for
roofing slates. In other places, they break into large slabs of
no great thickness, which have been used as flagstones.

Fossils are not abundant, except in some localities near the
base of the deposit. They are chiefly graptolites, the most
interesting being a species of *Monograptus* (*M. priodon*), *Cyrtograptus Murchisoni*, and *Retiolites Geinitzianus*. These are
often preserved in a state of high relief. They sometimes form
the nuclei of large elliptical concretions which contain a fair
proportion of calcareous matter.

The rocks shew indications of having been deposited in
water at some distance from the land; probably the area was
further from land than during the deposition of the Browgill
beds, but nearer to it than when the Skelgill beds were laid
down.

The original shaly character of the rocks is usually well exhibited
on somewhat weathered fragments of the slates. A series of very fine

4—2

lines marks the outcrop of the planes of lamination upon those of cleavage. The sediments shew a fairly uniform character from base to summit.

The frequency of the occurrence of elliptical concretions is of interest. They are often found, and occur abundantly in Brathay Quarry, west of the head of Windermere Lake, the quarry from which the division takes its name. They vary from a few inches to about a foot in the longer diameter, and, as stated, frequently contain organisms as nuclei; these are often graptolites, but occasionally other fossils play this part.

Similar elliptical concretions are found in beds of this age in other parts of the British Isles, and also abroad, as in Sweden, France and Bohemia. They are found in beds of other ages, but there must have been some original characters in the beds of Wenlock age which allowed of the frequent subsequent production of these concretions therein.

There is no sharp break between the Stockdale shales and the Brathay Flags. The uppermost Browgill beds have thin blue bands interstratified with the dominant green sediments. These blue bands become more frequent, and in the course of a few feet the green bands have disappeared, and the Brathay Flags assume their normal characters.

It was stated above that fossils are not abundant. This is true in the sense that determinable fossils are comparatively rare, but when one can break a portion of the highly cleaved rocks along a bedding plane, that plane is often covered with swarms of graptolites, which are so distorted that their specific and even in many cases their generic characters are quite unrecognisable. Most of the determinable fossils have been obtained from the lowest beds of the Brathay Flags. These beds have not been affected by subsequent cleavage to the same extent as the higher strata, having been protected from the effects of cleavage by the fairly massive grits of the underlying Browgill shales. In these lowest beds graptolites are most abundant, but corals, brachiopods, trilobites and cephalopods have also been found. In addition to the graptolites already mentioned, *Monograptus vomerinus* is abundant.

In the Cautley district near Sedbergh, the Brathay Flags have been much less affected by cleavage, and there graptolites are abundant throughout the deposits, and are sufficiently well preserved to allow of specific identification. In that district four graptolitic zones have been detected by Miss Watney and Miss Welch[1]. They occur in the following order:

Zone of *Cyrtograptus Lundgreni,*

„ *Cyrtograptus rigidus,*

„ *Monograptus riccartonensis,*

„ *Cyrtograptus Murchisoni.*

[1] Watney, G. R. and Welch, E. G., "The Zonal Classification of the Salopian Rocks of Cautley and Ravenstonedale," *Quarterly Journal of the Geological Society*, vol. LXVII (1911), p. 215.

I mention this, because there is little doubt that the same zones really occur in the Lake District, though for the reason given they have not yet been established.

Occasionally identifiable graptolites do occur some distance above the base. The basal deposits have furnished several specimens of *Cyrtograptus Murchisoni* and therefore that zone is readily recognisable in the district. I think that further investigation will shew that the rocks of some localities have locally escaped cleavage to a sufficient extent to allow of the detection of the higher zones also.

In any case it is perfectly clear that the Brathay Flags of the Lake District are the representatives of those of the Cautley district, and as the zones determined in the Cautley district are in all important respects identical with those which have been determined in the typical Wenlock strata of the Welsh Borderland, the equivalence of these Brathay Flags to the Wenlock strata is now definitely established.

The principal localities from which fossils of the *Cyrtograptus Murchisoni* zone have been obtained are Stockdale hamlet, the moorland on the west side of Troutbeck, Broughton Moor west of Coniston Lake, and the moorland behind Ireleth on the east side of the Duddon Valley.

Search should be made for fossils other than graptolites. Those already obtained shew that the fauna is of interest, and it is desirable that we should obtain further knowledge of its character.


# CHAPTER VIII

## LOWER PALAEOZOIC ROCKS

### SILURIAN SYSTEM

F.  *The Upper Coniston Flags, Coniston Grits and Bannisdale Slates.*

It is proposed to deal with these rocks together, as they represent the Lower Ludlow rocks of the Welsh Borderland. They are of great thickness, but the lithological characters indicate that they were deposited in comparatively shallow water, and no doubt accumulated rapidly. This is borne out by the fact that only two graptolite zones have been detected in them, as opposed to four zones in the much thinner Brathay Flags. It is true that five zones have been detected in the equivalent strata of the Welsh Borderland, and no doubt other zones than those already determined may be identified in Lakeland, but even then these very thick strata would roughly represent a period approximate to that of the formation of the Brathay Flags.

The strata of Lower Ludlow age begin immediately southsouth-west and south of the Brathay Flags, and except where the folds of the strata cause the development of newer rocks (the Kirkby Moor Flags) on the surface around Staveley, Kendal and the moorland near Kirkby Lonsdale, the deposits under consideration occupy all that part of the district south and east of the line mentioned, until one reaches the belt of Carboniferous rocks which occupy the country north of Morecambe Bay.

The prevalent dip of the beds is to the south-south-east, and south-east, and consequently the newer beds are generally found to the south and east of the older strata. An exception to this is found along the axis of a syncline extending from near Tebay to Staveley. Along this the Kirkby Moor Flags crop out at the surface, and older rocks (Bannisdale Slates) are found to the south of this. A much more important syncline, or rather basin, brings on the great development of Kirkby Moor Flags in the country between Kendal and Kirkby Lonsdale. This is outside the Lake District proper, but we shall have occasion to say something about this, the greatest development of the Kirkby Moor Flags, when we consider these strata in detail.

Apart from these major folds, the strata, and especially the Bannisdale Slates, are affected by countless minor folds which render the determination of their thicknesses a matter of some difficulty. The thicknesses assigned to the various divisions in the table on p. 7 must be regarded as merely approximate.

The outstanding feature, when we regard the lithological characters of the rocks of Lower Ludlow age, is the predominance of gritty material of the type known as "greywacke." It is most marked in the middle division, to which the name Coniston Grits has therefore been applied, but there is a considerable amount of gritty sediment among the strata of the Upper Coniston Flags, and still more among those of the Bannisdale Slates.

The Upper Coniston Flags form a fairly well-defined strip between the country south of Shap Wells and Broughton-in-Furness. On the east side of the Duddon Valley the outcrop owing to minor folding is less regular.

The beds have an approximate thickness of about 1500 feet, but it is probable that the highest beds are everywhere faulted out. They consist largely of laminated flaggy beds of a greyish colour weathering olive green, but they contain a much larger proportion of fine grit than do the underlying Brathay Flags; beds of somewhat coarser grit are not uncommon. The fauna is varied. Graptolites are abundant and are specifically different from those of the Brathay Flags. They are all *Monograpti*, and belong to a group of graptolites of which *Monograptus colonus* is the type: these are usually spoken of as the *coloniform* graptolites. Corals, crinoids, trilobites, brachiopods, lamellibranchs and cephalopods are not uncommon, and the fauna as a whole is richer than that of the Brathay Flags. Additions to this fauna are desirable. Most of the fossils hitherto found have been obtained from the moorlands on either side of Troutbeck, from the west side of Windermere, from the tract west of Coniston Lake, and from the district east of the Duddon estuary, but there is no reason why they should not be obtained from any place where these rocks are developed.

An interesting deposit occurs near the base. The lowest strata consist of rather coarse grits; they are succeeded by finer calcareous grits, which often stand out as a marked feature owing to their resistance to the weather. These calcareous beds contain *Cardiola interrupta*, a large number of cephalopods and some trilobites. On one bedding-plane which is traceable across miles of country are found swarms of the trilobite *Phacops obtusicaudatus*.

These Upper Coniston Flags are collectively known as the Coldwell Beds, from Coldwell Quarry, south of Brathay Quarry.

The Coniston Grits consist of some 4000 feet of gritstone, often of a fairly coarse type. Like the Coniston Flags, the outcrop forms a fairly straight strip between the east side of the district and the neighbourhood of Broughton-in-Furness, but is irregular on the east side of the Duddon estuary. About the centre of the grits is a band of flaggy mudstones usually not markedly cleaved. They were formerly quarried in Pennington's Quarry on the east side of Troutbeck, but are exposed in many other places.

Few fossils have been found in the grits of this division, but *coloniform* graptolites and some cephalopods have been obtained from the flaggy beds of Pennington's Quarry.

The Bannisdale Slates are marked by lithological characters which render them easily recognisable. They "may be described as sandy mudstones divided by thin bands of hard sandstone and occasional beds of grit. The sandy mudstones are much jointed and roughly cleaved, never making good slates, but often rough slabs, quarried for paving or building stones. The only tolerable slates were formerly worked in Bannisdale, in the very lowest beds of the series[1]." A thickness of over 5000 feet is calculated by Mr Aveline, but the difficulties of estimating thickness are especially great in the case of these beds. The rocks are usually of a leaden grey colour, the grits being often greenish, and the rapid alternation of thin grits and mudstones, of which there may be many in the thickness of an inch, gives the beds a curiously striped appearance which causes their easy recognition.

A few fossils occur in the Bannisdale Slates including the graptolite *Monograptus leintwardinensis* which is the highest zonal fossil of the Lower Ludlow strata.

The whole of the beds of this district which appertain to the Lower Ludlow age were obviously deposited rapidly, as indicated by the prevalence of Grit. No doubt oscillations occurred, marked by accumulation of alternating deposits formed with greater or less swiftness, as indicated for instance by the band of mudstone in the Coniston Grit. On the whole the evidence points to the Upper Coniston Flags having been accumulated less rapidly than the overlying beds.

---

[1] Aveline, W. T., *Memoirs of the Geological Survey*, "The Geology of the Country around Kendal, Sedbergh, Bowness and Tebay," second edition (1888), p. 17.

We may now give further particulars concerning these Lower Ludlow deposits. The writer has given an account of the various subdivisions in a paper in which fossil-lists of the principal subdivisions will be found[1].

The Upper Coniston Flags have been divided into Lower, Middle and Upper Coldwell beds. Of these the two lower divisions are thin, and the great mass of the deposit belongs to the upper division.

The Brathay Flags are succeeded by massive grits,—the Lower Coldwell beds. These have hitherto yielded no fossils, but are grouped with the Upper rather than the Lower Coniston Flags on lithological grounds.

The Middle Coldwell beds have been sufficiently described so far as their lithological character is concerned. As stated, they usually stand out as a marked ridge. This is specially noticeable on the east side of Troutbeck, where the ridge runs up Applethwaite Common and is seen as a small peaked elevation against the sky line at the summit.

The fossils are of interest. As already stated, *Phacops obtusicaudatus* appears to be confined to a single bedding plane, where it occurs in swarms. This plane is easily detected, for about a foot above it is another bed with abundant *Orthocerata*. These are the most prominent bedding planes in the series. As the beds of the division furnish good flagstones, numerous small quarries have been excavated along the outcrop. The principal are on the east side of Troutbeck above and below the Garbourn road; near Skelgill, in a field just below the moorland lane on the left side of the valley, where another lane leaves it for High Skelgill Farm; and at Coldwell, west of Windermere Lake. In addition to the fossils named, *Cardiola interrupta* is fairly abundant. The whole fauna presents an unmistakeable Lower Ludlow facies.

The Upper Coldwell beds contain several fossils which occur in the middle division, including *Cardiola interrupta* and certain species of *Orthoceras*. In addition we find graptolites including *Monograptus bohemicus*, *M. rœmeri* and *coloniform* species.

According to Miss Watney and Miss Welch[2] the Middle Coldwell beds may represent the zone of *Monograptus vulgaris* in the type-district of the Welsh Borders, and the Upper Coldwell beds certainly represent, in part at any rate, the succeeding zone, that of *M. Nilssoni*. Among other fossils of the Upper Coldwell beds we may notice *Actinocrinus?* *pulcher*, *Ceratiocaris Murchisoni* and *Phacops Stokesii*. The chief fossil localities are the east side of Troutbeck, the west side of the same valley between it and Skelgill, a quarry south of Coldwell Quarry, and various places west of Coniston Lake, but no doubt many others will

---

[1] Marr, J. E., "On the Wenlock and Ludlow strata of the Lake District," *Geological Magazine*, Decade III, vol. IX, p. 534.

[2] *loc. cit.*

be discovered, for the beds seem to yield some fossils in nearly all localities.

It has been stated that the highest beds of the Upper Coniston Flags are probably everywhere concealed owing to faulting. The nature of this fault will be discussed later. It is possible that higher graptolitic zones may be concealed in this way, for the two zones of the Welsh Borders—those of *Monograptus scanicus* and *M. tumescens*—have not been detected in the Lake District, though the Coniston Grits probably represent at least one of them.

It has been stated that these Coniston Grits have a band of argillaceous rock about the centre. The rocks of this band are locally termed "sheerbate flags," for they break along the bedding. They yield *coloniform* graptolites, but those hitherto found are not sufficiently well preserved to enable us to determine their zonal horizon.

Professor Hughes has described, under the name of "Crook and Winder grit," a coarse band of calcareous grit which is found among the Coniston Grits of the hills near Sedbergh. This grit may also be represented in the Lake District, and should be looked for. A specimen of yellow grit collected by Ruthven and preserved in the Kendal Museum is labelled "Applethwaite." It contains *Monograpti*, crinoids and brachiopods, and may be representative of the Winder grit.

Since the above was written, the Rev. Canon Crewdson has discovered a similar grit at Calgarth on the eastern shores of Windermere, and little more than a mile distant from Applethwaite[1]. It is perhaps on a somewhat different horizon from the Winder grit but is of the same general age. It is, as one would expect from the characters of the other strata, somewhat finer in grain than is the Winder grit, but otherwise is lithologically similar. It has furnished a number of fossils of Lower Ludlow age, which have been determined by Miss G. L. Elles, D.Sc. Among the more important of these are:

*Skenidium Lewisii, Cœlospira hemispherica, Camarotaechia nucula, Dayia* cf. *navicula, Zygospira* sp., and *Holopella* cf. *tenuicincta.*

The Bannisdale Slates are not very fossiliferous in the district, though many fossils have been found in them among the Howgill Fells. A few brachiopods and *Orthocerata* have been discovered and occasional fragments of trilobites. Graptolites are rare in Lakeland, though abundant in places in the Howgill Fells. Some have been found at Reston near Ings, between Windermere village and Staveley. Fortunately they belong to a definite zonal species, *Monograptus leintwardinensis*, which, as already stated, characterises the highest graptolitic zone of the Lower Ludlow rocks.

At Tottlebank Farm, about two miles south-west of the foot of

[1] Crewdson, G., "Coniston Grits of Windermere," *Geological Magazine,* Decade VI, vol. II (1915), p. 169.

Coniston Lake, a limestone occurs in the Bannisdale Slates. It was discovered by Sedgwick, who records its presence on p. 219 of his *Letters* on the Geology of the Lake District, appended to the third edition of Wordsworth's *Guide*. It should be carefully examined for fossils.

The conditions under which these Silurian rocks were accumulated have already been stated in general terms, but we shall have more to say upon this point after describing the representatives of the Upper Ludlow rocks in our district.

# CHAPTER IX

## LOWER PALAEOZOIC ROCKS

### SILURIAN SYSTEM

#### G. *The Kirkby Moor Flags.*

The highest division of the Lower Palaeozoic rocks of Lakeland has long been celebrated for the abundance and variety of its fossils. The rich collections were largely due to the work of local geologists around Kendal, prominent among whom were John Ruthven, Rev. T. B. Danby, and Messrs Gough, Sharpe, and Marshall.

The distribution of these beds has already been given. The most extensive development is that lying outside the borders of the Lake District proper. The beds are found in a triangular tract with its base east of Kendal and its apex near Kirkby Lonsdale. The tract is about 10 miles in length from north to south and seven miles along the base at its northern boundary, which extends from a point just north of Kendal to another a little west of the river Lune. The highest beds of the series in this tract are found near Kirkby Lonsdale. A small but very important development is found to the west of this, being separated from it by the Carboniferous rocks of Kendal Fell. This little patch is of much importance, as it exhibits certain beds of an intermediate date between the Bannisdale Slates and the main development of the Kirkby

Moor Flags. The beds of this patch are well shewn near the village of Underbarrow. The third tract, as already stated, is situated in the centre of a synclinal fold. The main development along this fold is between the village of Ings and Borrow Beck, but a small outlying patch is also found west of Tebay. It is not unlikely that the Kirkby Moor Flags occur further west along this synclinal fold. Rocks resembling them are found on the west side of Windermere to the north of Far Sawrey, where they contain brachiopods and other fossils. The total thickness of the beds in the large tract is about 1500 feet.

The strata are usually recognised with ease. They consist of grey-green flagstones, containing a considerable proportion of gritty matter, with subordinate bands of coarser grit. A good deal of mica is often present, and shews up on the bedding-planes. A certain amount of calcareous matter usually occurs and is specially abundant along certain bands which are rich in fossils. These bands "when decomposed, form conspicuous lines of soft brown earthy rock," which form a marked feature.

The deposits near Underbarrow, which are mentioned above, present lithological characters to some extent intermediate between those of the Bannisdale Slates, and of the higher part of the Kirkby Moor Flags. They contain many fossils including the brachiopod *Rhynchonella (Dayia) navicula*, which often occurs in abundance along definite bands. An interesting feature of these beds is the occurrence of a calcareous band containing fossil starfish. It is known as the "Starfish-band." A description of these beds will be found in Professor Sedgwick's *Letters* on the Geology of the District in Wordsworth's *Description of the Scenery* (Supplementary Letter, IV). Similar starfish have been found at Potter Fell and Shepherd's Quarry in the Kendal district, and should be searched for elsewhere at the base of the Kirkby Moor Flags.

The fauna of the general mass of the Kirkby Moor Flags is a varied one. The fossiliferous bands which weather to a brown earthy rock are often composed largely of the gastropod shell *Holopella gregaria*. Among the brachiopods *Rhynchonella (Camarotaechia) nucula* and *Chonetes striatella* are abundant. Lamellibranchs, often of large size, are fairly frequent. They were peculiarly abundant in flaggy beds on Benson Knott near Kendal, which beds are now no longer exposed. The highest beds seen near Kirkby Lonsdale are also flaggy, with numerous lamellibranch shells and other fossils.

The fauna is clearly that of the Upper Ludlow rocks of the Welsh Borders, but the newest Lower Palaeozoic rocks of Lakeland do not represent the very highest subdivision of this series in the borderland. These may once have been deposited, and removed by subsequent erosion, but there is no evidence of this.

Lists of fossils of the Kirkby Moor Flags are found in many publications. A list is given in Prof. Sedgwick's letters. Another is found in the Geological Survey Memoir of the Country around Kendal, Sedbergh, Bowness and Tebay to which reference has been made, and yet another in the Survey Memoir of the district south of this[1].

The passage-beds which occur at Underbarrow and elsewhere will certainly repay further study. I give here a condensed account of Prof. Sedgwick's description of these in the letter referred to above. They are described in ascending order:

Beginning with the faulted beds of the valley near Underbarrow Chapel we find on the line of the Kendal road:

"(a) A thick group of coarse slate and flagstone...extending nearly to a farm called High Thorns...it contains *Terebratula* [= *Dayia*] *navicula*....

"(b) A bed six or seven feet thick with two species of *Asterias* [three species are now known, namely *Palaeaster hirudo*, *P. Ruthveni*, and *Palaeasterina primaeva*].

"(c) Flags...(without cleavage): some red calcareous bands with many fossils. Numerous *Trilobites*....

"(d) ...Hard grits...and flags...identical with those under Benson Knott." These are the normal deposits of the Kirkby Moor Flags.

The ordinary upper deposits of the Kirkby Moor Flags require no further description.

Thanks to the work of Miss Elles, D.Sc. and Miss Slater, we now know the detailed succession of the Ludlow rocks of the Ludlow district[2]. These authors divide the rocks of that district into three divisions, namely the Aymestry group at the base, the Upper Ludlow group in the middle and the Temeside group at the summit. The Aymestry group has the Aymestry limestone with *Conchidium Knightii* below, and shales with *Dayia navicula* above. No recognisable equivalent of the

[1] "Geology of the neighbourhood of Kirkby Lonsdale and Kendal" (1872)

[2] Elles, G. L. and Slater, I. L., "The highest Silurian Rocks of the Ludlow District," *Quarterly Journal of the Geological Society* (1906), vol. LXII, p. 195.

*Conchidium* limestone has been detected in Lakeland but the beds (*a*), (*b*) and (*c*) of Sedgwick's succession near Underbarrow may well represent the *Dayia*-shales.

The bulk of the Kirkby Moor Flags is certainly the equivalent of the Upper Ludlow group as a whole or in part. This group is divided into a lower zone with *Rhynchonella nucula* and an upper zone with *Chonetes striatella*. These fossils, as seen, are very common in the Kirkby Moor Flags. It remains to be discovered whether their distribution therein will enable us to differentiate these zones in Lakeland. It is doubtful whether any representative of the Temeside or *Eurypterus*-shales is present at the summit of the Kirkby Moor Flags. *Eurypterus cephalaspis* is recorded from Kirkby Moor and preserved in the Sedgwick Museum, Cambridge, but the discovery of one Eurypterid is not sufficient to prove that we are dealing with representatives of these *Eurypterus*-shales of the Ludlow district.

# CHAPTER X

## DISCUSSION OF THE CONDITIONS UNDER WHICH THE ROCKS OF LOWER PALAEOZOIC AGE WERE FORMED IN LAKELAND

Some indication of the conditions under which the different members of the Lower Palaeozoic succession were accumulated has been given in the preceding chapters, but it is desirable that we should get a more definite idea of the physical conditions which prevailed in the area during their formation.

The characters of the sediments of the Lake District and of the southern part of Scotland are explicable upon the view that a continental tract lay to the north of Scotland and the adjoining regions through Lower Palaeozoic times, and that there was on the whole a gradual southerly encroachment of the coast-line of this continent in the later part of this period, so that, with minor variations, the sediments of later series are of a shallower water character than those of the earlier, for they were laid down nearer to the coastal margin.

The Lake District sediments also suggest, through the nature of their local variations, that on the whole deeper water lay towards the west of the district than along its eastern border, which would be explicable on the supposition that a tract of land of some size lay to the east, during some periods, perhaps on the site of what are now the Pennine Hills.

As a minor feature, we must note the occurrence of marked shallow-water conditions during the accumulation of the rocks of Caradocian age. These conditions may have been the result of local earth-movements, giving rise to temporary land-masses in the vicinity of the district, or a considerable shallowing of the ocean-waters may have resulted from the piling up of volcanic material during the accumulation of the rocks of the Borrowdale Series. The latter seems, with our present knowledge, to be the most likely explanation, but the question is bound up with the actual age of the rocks of the Borrowdale Series. If these be, as at present generally stated, of Ordovician date, their accumulation must have had some effect upon the depth of the Ordovician sea, but if, as has been suggested, the Borrowdale rocks are of a date anterior to that of the deposition of the Skiddaw Slates, their eruption can have produced no direct effect upon the shallowing of the seas in which the earlier deposits of the Coniston Limestone were laid down.

Beginning with the Skiddaw Slates, we have seen that a large proportion of the deposits gives indications of having been formed in seas of no great depth. It is doubtful whether any rocks equivalent to the older portions of the Skiddaw Slates occur in the southern part of Scotland, but the general shallow-water characters of these slates suggest their formation in sea-tracts at no great distance from a coast-line. No doubt many oscillations occurred during the formation of this complex and apparently thick group of sediments, which, as has been seen, may be representatives of more than one geological period.

No physical break of the nature of an unconformity has been detected in the sediments; but there is at any rate a suggestion of one below the "Skiddaw Grit" which is sometimes conglomeratic, containing pebbles of earlier rocks. This of course might imply merely local erosion, and not general uplift. Until more detailed work has been accomplished among these slates, the consideration of their origin must be left thus vague.

There is indication of a deepening of the sea-floor when the Ellergill division of the Upper Skiddaw Slates was laid down. These deposits, both in the district and elsewhere, consist of much finer muds than do the earlier divisions. The black earthy mudstones are very similar in character in Lakeland, the east side of the Eden Valley, the southern uplands of Scotland and Wales: nay, a similar type is found in some tracts of the Continent also. This wide distribution of fine mud containing a rich graptolite fauna with few other organic remains almost certainly points to open-sea conditions in the tracts where the mud was deposited.

We have seen that, during the formation of the upper (Milburn) division of the Upper Skiddaw Slates, volcanic activity set in, resulting in the outpouring of andesitic lavas and the showering forth of ashes of a similar composition. These volcanic outpourings were no doubt submarine, for they are intercalated with normal sediments containing marine fossils. According to the ordinary interpretation, these rocks marked the beginning of the great period of volcanic activity which culminated in Llandeilo times, and caused the accumulation of the great Borrowdale Series.

Two questions are suggested in connexion with the Borrowdale rocks : firstly, were they as a whole of subaerial or submarine origin ; secondly, what was the nature of the vents from which the materials were emitted ? The first question is a difficult one. Clifton Ward was certainly in favour of the view that the main mass was of subaerial origin, and this is suggested by the entire absence of fossils. He supposed that the earlier outpourings were submarine, a view which receives apparent support from the occurrence of sediments of lithological characters like those of the Skiddaw Slates, which are intercalated among the lowest volcanic rocks seen in Cat Gill near Falcon Crag. Ward also supposed that the volcanoes soon appeared above the sea, partly owing to upheaval of the sea-floor, and partly to the actual filling up by volcanic material. In the rocks of this series above those of Cat Gill we have hitherto met with nothing resembling marine sediments until the rocks of the Scawfell ash and breccia group are reached, in which is the limestone near Hole Rake, Coniston, to which reference has been made. This may be a marine sediment, and probably belongs to the rocks of the Borrowdale Series ; but neither of these views can be regarded as established. A microscopic examination of the limestone gives no evidence of organic structure.

Turning now to the nature of the vents from which the material was extruded, Ward regarded the rocks as having been forced up ordinary volcanic pipes subsequently filled with intrusive rocks to form volcanic necks. One of these he believed to be represented by the igneous rock of Castle Head, Keswick, and he alludes to other igneous masses which may have had a similar origin. Sir Archibald Geikie in his *Ancient Volcanoes of Great Britain* also calls attention to the existence of coarse breccias which may have filled similar necks, and suggests that "if any one great volcano existed, its site must lie outside of the present volcanic district, or more probably, that many scattered vents threw out their lavas and ashes over no very wide area, but near enough to each other to allow their ejected materials to meet and mingle." We have seen reason for believing that much of the igneous material was not brought to the surface in the form of lava-flows and ashes, but was forced along planes of weakness of the Skiddaw Slates and volcanic rocks in the form of laccolithic masses, sills and dykes. The molten rock from some of the latter may well have reached the surface and given rise to fissure eruptions. In this case the rocks of the series may have been brought through apertures of different kinds, and both fissure-eruptions and eruptions through isolated vents may have contributed to the building up of a complex plateau, such as Dr Thoroddsen has proved to have taken place in the case of Iceland and Sir Archibald Geikie in that of the Tertiary volcanic rocks of the west of Scotland.

Turning now to the conditions which prevailed before the deposition of the Coniston Limestone, we meet with another difficulty. It has been stated that one view is, that the rocks of this group repose with a very marked unconformity upon the rocks of the underlying Borrowdale Series, and that this unconformity is most marked at the south-western end of the district. According to this view the volcanic rocks must have been folded, and have undergone much erosion before the deposition of the Coniston Limestone. Whatever be the nature of the junction, all are agreed that the hiatus is less to the east of the district, near Shap, than at its south-western extremity. It is in the eastern part that the rhyolitic lavas at the top of the Borrowdale Series are best developed, and that rhyolitic rocks of similar characters are actually intercalated between the marine sediments of the Coniston Limestone.

If these two masses of rhyolitic rock are genetically related, it would seem that the hiatus here is not great, and that just as we appear to have intermittent volcanic activity at the end of Skiddaw Slate times ushering in the great period of eruption represented by the Borrowdale Series, so at the summit we find similar intermittent activity marking the final dying out of this great volcanic phase.

The Coniston Limestone deposits were, as a whole, laid down in shallow water, which, according to one view, is due to the filling in of the sea by volcanic materials and subsequent depression, causing shallow water, and to the other owing to uplift producing unconformity. The two causes may of course have combined. In any case the nature of the sediments is such as to shew that they were deposited in water of no great depth, and the characters of the fossils bear out this conclusion. Also, whether or no there is an unconformity at the base of the Coniston Limestone, there is certainly one, though of no great importance, in the series itself, as indicated by the conglomerate at the horizon above the rhyolitic group of Yarlside and Stockdale.

In the Ashgillian period the water appears to have become deeper, as shewn by the fairly pure limestones and fine mudstones of this series. The development of these lithological types far afield from Lakeland indicates that the change was no local one, but probably affected a large part of north-western Europe.

We now reach the plane of demarcation between the Ordovician and Silurian rocks. It has been seen that the evidence points to a conformable junction of the strata of the two systems in Lakeland, but there is no evidence of the gradual passage of one type of sediment into the other. On the contrary, the plane of junction is marked by sharp contrast. It has been stated also that a palaeontological break of some importance occurs at this horizon.   Perhaps the most likely view is that there was a pause in the accumulation of sediment at this stage, which would account for the phenomena just mentioned.

Before discussing in detail the physical conditions which prevailed in Silurian times, we must quit the district for a moment to consider the information supplied by the rocks of the southern uplands of Scotland. The detailed researches by Prof. Lapworth in that region have brought out in a prominent manner the fact, already mentioned, of a gradual southerly travelling of the coast line of the old Caledonian continental tract.

The representatives of the Llandeilo, Caradoc, Ashgill and Llandovery strata, which have a thickness of almost 4000 feet at the northern part of the southern uplands, thin out southward, so that their united thickness about Moffat is only 300 to 400 feet, and the sediments of the north indicate shallower water conditions than in the south.   The encroachment of the shore-line is indicated by the extension of shallow-water deposits to the Moffat area in Gala-Tarannon times, and in the Lake District itself, the equivalents of the Scottish Gala beds—the Browgill Shales— shew much shallower water conditions in that area than do the underlying Skelgill beds.   The accumulation of the Wenlock beds was apparently marked by a pause in the encroachment, and even by a temporary recession of the shore-line in a northerly direction, but in Ludlow times the southerly extension of the line became very marked, and is indicated in the Lake District by the gritty characters of the Ludlow rocks, and ultimately by the conversion of the Lake District area into land and the shifting of the coast-line to the south of the district.

We may now consider the effects of the suggested land tract to the east of the district.   The higher Ordovician strata (Caradocian-Ashgillian), which when volcanic rocks are absent have a thickness of about 250 feet in the district, swell out to many hundreds of feet of more purely mechanical sediment about Sedbergh.   The Silurian strata also become thicker and more coarse-grained in the Sedbergh area.   This thickening was proved by Prof. Nicholson and myself in the case of the two divisions of the Stockdale Shales.   It is not apparent in strata of Wenlock age, but becomes marked in those of Lower Ludlow date: (the Upper Ludlow rocks are not represented in the Sedbergh area).

It is unnecessary to enter into any further details concerning the changes in the lithology of the Silurian rocks and their significance. These matters have been discussed in the preceding chapters.

# CHAPTER XI

## THE CHANGES AT THE END OF LOWER
## PALAEOZOIC TIMES

At the close of Silurian times, a great change came over the district. The Lower Palaeozoic sea in which the sediments which we have described were deposited was converted into land, and with one doubtful and insignificant exception no deposits were formed in the area during the Devonian period. The rocks next in order to the Silurian rocks of Lakeland are of Carboniferous age. The changes which converted sea into land produced profound effects upon the characters of the rocks which were formed previously, and it is these effects which we must now consider.

The Lower Palaeozoic rocks underwent a great uplift, using this term in a popular sense to indicate what we may regard as an elevation of the sea-floor, in this case to such an extent that this sea-floor was converted into land. One result of this movement was that the strata originally deposited in horizontal sheets became inclined at various angles to the horizon.

The age of the uplift is easily determined. The Lower Palaeozoic strata are inclined as an effect of the movement, while the Carboniferous rocks are not affected by it. This is of course because the latter were not then in existence, and accordingly the date of the movement is fixed as post-Silurian and pre-Carboniferous, in other words, it is of Devonian age.

One important effect of the movement upon the Lower Palaeozoic rocks was briefly noticed in the introductory chapter. It was there stated that the beds were folded into an arch, along a line having a general east-north-east and west-south-west trend, which passed through Skiddaw. The section (Fig. 1) illustrates this, and that section at its north and south ends shews the Carboniferous rocks resting unconformably upon the upturned edges of the strata folded into this arch. The top of the arch was swept away by erosion in the period prior to the deposition of the Carboniferous rocks. The nature and effects of this erosion will presently be considered.

The great fold above mentioned was itself only a local effect of a set of movements which profoundly affected the Lower Palaeozoic rocks over large areas of Britain and north-western Europe, and it is further complicated by minor folds of varying degrees of importance. The occurrence of three synclines or troughs has already been noted, one of small size among the Borrowdale rocks of the fells between Borrowdale and the Vale of St John, another of more importance in the same rocks, the axis of which passes through the Scawfell group of hills, and a third in the Silurian rocks which causes the outcrop of the Kirkby Moor Flags along the Staveley-Tebay tract, with older rocks on either side. Between these synclinal folds are of course corresponding anticlines or saddles.

Still smaller folds are met with in profusion. Some of them are many scores of yards across when measured at right angles to the lines of the axes of the folds. Others are only a few yards or even feet across, others again are measured by inches, and the smallest are on so minute a scale that we may find several of them in a piece of rock an inch across.

These folds were the result of pressure acting on the rocks. Such folds may be produced by lateral compression acting in opposite directions, when the folds will arise with axes at right angles to the directions of the application of pressure. But a similar result may be brought about by pressure exerted in one direction, if the rocks be pressed against a barrier, such as would be formed by a mass of hard rock lying in the direction away from that whence the pressure comes. There is reason to suppose that this was the case in the Lake District, and as the axes of the folds have a trend not far removed from an east and west direction the pressure must have come either from the north or from the south; at present we know not which: there is some evidence that the movement came from a southward direction, but into this point we need not here enter.

Rocks which are affected by such pressure do not necessarily adapt themselves to the changed conditions by folding only.

They are frequently broken across, and blocks on different sides of the planes of fracture become displaced in various degrees. Thus a set of faults is developed subsidiary to the folds, and the nature of these faults must now be considered.

The evidence of a fault between the Skiddaw Slates and the Borrowdale Series along nearly the whole length of their line of junction has already been noticed. This fault is inclined at a low angle to the horizontal, so that one set of rocks has moved further forward than the other in a nearly horizontal direction along this fault-plane.

In many places this fault is accompanied by a breccia consisting of broken angular fragments of Skiddaw Slate and volcanic rock in a fine-grained ground-mass. Such intermingling of rocks of two ages suggests that the rock is a fault-breccia and not one due to volcanic explosions. Similar breccias are found at various horizons among the rocks of the Borrowdale Series, and suggest the occurrence of faults at those horizons: that such faults actually do occur is known on other evidence.

The possible existence of a similar fault at the top of the Borrowdale Series separating those rocks from the beds of the Coniston Limestone has already been spoken of. Various other faults of this type are found among the Silurian rocks; one in particular separates the Coniston Flags from the Coniston Grits. Several of the faults which affect the rocks of the district have fissures which are vertical or closely approximate to the vertical position. Many of these are ordinary faults, where the rocks have been vertically displaced, causing the upthrow and downthrow sides characteristic of such faults.

Along other of these vertical fissures there is evidence that movement has gone on in a direction approximately horizontal, which may be illustrated by taking two books and pushing one sideways along the other. That movements of this nature had taken place in the district was suspected by Prof. Sedgwick so long ago as 1831[1]. He remarks: "the usual appearance on the opposite sides of the faults above described, is exactly that of a great horizontal lateral movement, and is not, I believe, by

[1] Sedgwick, A., "Introduction to the general structure of the Cumbrian Mountains," *Transactions of the Geological Society*, 2nd series, vol. IV, p. 47.

any means entirely deceptive; for expansive forces of elevation acting on oblique planes might easily produce such a movement. The effect was probably of a compound kind."

These movements are best illustrated where the faults affect the Coniston Limestone. The limestone is in some cases shifted horizontally for more than a mile from its original position. Apparent horizontal shifting is of course characteristic of vertical movement of inclined strata, when the latter have been eroded

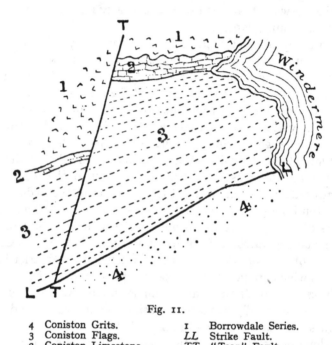

Fig. 11.

| | | | |
|---|---|---|---|
| 4 | Coniston Grits. | 1 | Borrowdale Series. |
| 3 | Coniston Flags. | LL | Strike Fault. |
| 2 | Coniston Limestone. | TT | "Tear" Fault. |

to a level surface. In the cases under consideration, however, this lateral shift only affects thin strips of strata. The faults which affect the Coniston Limestone do not affect the Coniston Grits to the south, nor do they extend far into the rocks of the Borrowdale Series to the north. They are limited to slices of rock situated between two of the horizontal faults having nearly horizontal fissures, and indicate that the slices between these fissures not only moved at different rates, but that portions of one slice were also moving at different speeds, and that the faster moving portion was torn away from the one with slower motion. Faults of this character have therefore been spoken of as *tear-faults*.

The plan in Fig. 11 shews one of these tear-faults which affects the Coniston Limestone to the west of the head of Windermere. It is seen ending abruptly against the nearly horizontal fault (termed *strike-fault* in the figure) on the south side. It also ends in the same way against a similar fault in the volcanic rocks to the north.

The effect of these tear-faults is frequently marked by the production of a belt of shattered rock of brecciated character, possessing a considerable width, along the line of movement. A belt of this nature may also be produced when unaccompanied by permanent shifting of the rocks, for these may, after backward and forward movement of sufficient amount to produce brecciation, be brought back to their original position. I have spoken of such belts as "shatter-belts." They are of importance incidentally as causing tracts of weakness which may readily be affected by erosion.

These shatter-belts are widely distributed throughout the district. Their nature may be best studied among the flinty rocks of the Scawfell Ash and Breccia group on the Scawfell hills, where they are markedly contrasted with the durable flinty rocks. The belts are often many yards in width. The rock is broken into a rubbly mass of angular fragments of varying size in a matrix of finer crushed material. In the Scawfell hills subsequent introduction of an iron compound has produced a red coloration of the rocks of the belt, which enables them to be readily discriminated, when seen against the greenish flinty rocks through which they cut. They are admirably shewn along a line between Sty Head Tarn and the foot of Rossett Gill, and again in a line at right angles to this, in Ruddy Gill to the east of Sprinkling Tarn.

Three important shatter-belts radiate from a tract of ground in Great Langdale. One extends westward through Little Langdale, over Wrynose Pass, across the Duddon Valley, over Hardknott Pass, along part of the Eskdale Valley to Eskdale Green, and then over a depression to Miterdale. This belt has an east-north-easterly to west-south-westerly trend, and is traceable for about 15 miles. The second belt runs northward from Great Langdale through a depression in Red Bank to Grasmere, thence over Dunmail Raise, down the Thirlmere and Naddle Valleys, and probably still northward through the Skiddaw Slates of the Glenderaterra Valley, a total distance of nearly 20 miles.

Fig. 12. Head of Ennerdale.

Shewing position of shatter-belt between Great Gable (on right) and Green Gable (on left).

The third belt leaves the first about four miles west of the point where the second diverges. It is traceable from near Little Langdale Tarn north-westward into Great Langdale, thence up the Mickleden branch, over Rossett Gill, past Angle, Sprinkling and Sty Head Tarns, through Windy Gap, between Great and Green Gable, and thence down the head of the Ennerdale Valley. Its total length is about 11 miles. Fig. 12

shews this belt at the head of Ennerdale. The gap on the sky-line and the gully extending to the right from this gap are caused by it.

It is interesting to notice that these belts diverge in that part of the district where the strike of the Lower Palaeozoic rocks changes from an east-north-east—west-south-west direction to one ranging from north-east to south-west.

The situation of other shatter-belts will be noticed when we consider their influence upon the operations of erosion.

As lateral compression of the rocks proceeded, the time came when the rocks were no longer able to adapt themselves to the movements by further folding and faulting. It can readily be seen by taking strips of dough to represent strata that folding of the dough cannot go on indefinitely. When it has been folded to its utmost extent, if lateral pressure be still exerted, the dough will adapt itself by change of its particles which will cause the mass to become narrower in a horizontal direction and to rise in height. The rocks of the earth's crust are similarly affected. After folding has occurred to its utmost extent, the particles of the rocks undergo rearrangement. Not only are their longer axes arranged at right angles to the direction of pressure, but the particles are themselves flattened. Owing to this the rocks now possess a fissility along planes at right angles to the direction of pressure. They are affected by slaty cleavage and converted into slates. This cleavage affects rocks of all the Lower Palaeozoic divisions, when they are of such a character as to take on the cleavage structure. The massive grits usually escape, as do also some of the lavas and coarse volcanic agglomerates, but even these volcanic rocks have been affected when the pressure has been intense.

The Skiddaw Slates usually shew imperfect cleavage, and accordingly we find no good roofing-slates among the rocks of this division. It has been noted that the Skiddaw Slates to the north of the district are only slightly affected by cleavage. The volcanic rocks yield good roofing-slates. These slates, as already noticed, run along two definite belts. The northerly one, of no great extent, extends from the east side of Borrowdale to Honister. The southern belt is traceable from

Mosedale, west of Shap, to the district west of Coniston Lake, and subsidiary bands which yield slates are found in the neighbourhood of this line, being worked in the Duddon Valley.

The northern belt, as already stated, runs obliquely to the strata and passes from the rocks of the Ullswater group to those of the Falcon Crag group. Fine ashes are affected by cleavage at Honister, and furnish the best slates, but in Borrowdale slaty rocks are found in this belt consisting of lavas and volcanic breccias.

The southern slate band is essentially confined to one division, that of the Scawfell ash and breccia group, and the fine ashes in this band form excellent slates which have been worked at various points along the band: even here however we meet with well cleaved breccias and occasionally lavas.

That the cleavage was impressed after the occurrence of folding and faulting is shewn by numerous examples of folds and faults which are affected by the cleavage, the fault-planes being often sealed up by cleavage so that the rocks no longer break with ease along those planes. This often occurs on a small scale. Beautiful miniature examples of such folds are found among the slates of Walney Scar, and the small faults of the Tilberthwaite slates have long been celebrated.

In addition to the cleavage, which may be regarded as a metamorphic effect of mechanical nature, a certain amount of mineral change was also set up in the rocks as the result of pressure. How far such change has taken place in the body of the rocks can only be ultimately settled when a much fuller microscopic examination of these rocks has been carried out than has already been made.

To this point we shall presently return. An examination of many rocks with the unaided eye shews however that some of the components have undergone alteration.

In many of the highly compressed Skiddaw Slates the divisional planes are marked by a sheeny surface composed of some form of mica, and it is clear from its conditions of occurrence that the mica is of secondary origin, having been derived from some of the constituents of the rock.

In many of the rocks of the Borrowdale Series interesting changes have occurred. Where porphyritic lavas have undergone great compression, not only do we find the development of abundant mica on the cleavage surfaces, but the felspars have been flattened and have obviously undergone mineralogical change, with the production of white mica. In the case of the garnet-bearing rocks in similar circumstances, the garnets have been partly or entirely converted into white mica, or even into chlorite.

There is little doubt that crystallisation of the purer limestones of the Coniston Limestone Series is also due to earth-stresses, the limestones having crystallised where there has been relief from pressure. It is probable that the crystalline character of the great mass of white limestone in the neighbourhood of Millom is due to this cause.

In the Silurian rocks the mineralogical changes produced by pressure are not so apparent, but in addition to the large amount of detrital mica which is present in the rocks of the Bannisdale Slates, there is much of this mineral which appears to be of secondary origin.

We may now consider the nature of the folding, faulting, cleavage and metamorphism in greater detail.

As has already been stated, the ordinary view is that the arch through Skiddaw is the most important tectonic structure exhibited by the Lower Palaeozoic rocks, but more than one suggestion has been made that faulting and not folding is in reality more important.

Geologists have long regarded the possibility of the rocks of the Borrowdale Series being older than those of the Skiddaw Slates, and thrust over the latter along the great boundary-fault, which would in that case be an overthrust-fault. This view was expressed by Prof. Lapworth in a lecture delivered at Keswick in 1883.

From a report in a local paper I extract the following: "Professor Lapworth initiated a theory in reference to the comparative ages of the different formations in this district, quite heretical to what has hitherto been held. Notwithstanding that the Skiddaw slate appears to underlie the lavas, ashes, and porphyries, he had formed the opinion that the position was accidental....He believed that the hard rocks [the lavas etc.] were the earlier formations."

Mr J. G. Goodchild also raised the question as to whether the Borrowdale Series might not really be older than the Skiddaw Slates, though the facts which he took into consideration appeared to him to militate against the hypothesis[1].

The researches of Dr Harker and myself led us to suggest that the fault at the base of the Borrowdale Series is not an overthrust, but what we termed a *lag-fault*, the upper and newer Borrowdale rocks having lagged behind the older Skiddaw Slates, owing to slower movement than that of the older rocks. Such a lagging would necessitate a thrust-plane somewhere in advance of the lagging mass, and it was suggested that a thrust of this nature existed, and that traces of it might actually be detected in the district itself. We inferred that the tear-faults which have already been mentioned, were subsidiary to the thrust and lag-planes[2].

A third possibility which must be considered is that the fault between the two groups is a thrust-plane and not a lag-plane, in that the upper rocks have been pushed further than the lower and not lagged behind, but that the upper rocks are nevertheless newer than the lower and not older.

The nature of the evidence as bearing upon these three views must now be considered.

We will first take into account the possibility of inversion along the great fault.

It is quite clear that the fault-plane is one departing not far from the horizontal. This is shewn by the outcrop of the fault along the line of country which is represented on the Geological Survey Map of the Keswick district on the scale of one mile to the inch. The fault is seen to crop out in a series of zigzags, the apices of these being directed alternately northward and southward. They point to the north when crossing ridges, the actual apex being at the ridge summit, and to the south when crossing valleys, the apex in this case being at the valley bottom. Clifton Ward supposed that we were here dealing with cross-faults nearly at right angles to one another, but that would not account for the constant coincidence of the apices with ridge summits and valley bases respectively. Detailed mapping certainly shews that we are dealing with one fault, and that the nature of the outcrop is due to its gentle inclination from the horizontal plane. Along this very line Mr Dakyns long ago detected the gentle inclination of the plane of junction, though he attributed it to an unconformity and not to a fault[3]. He estimates the angle of junction as not more than 30°.

---

[1] Goodchild, J. G., "Observations upon the Stratigraphical Relations of the Skiddaw Slates," *Proceedings of the Geologists' Association*, vol. IX, p. 469.

[2] See Marr, J. E., "Notes on the Geology of the English Lake District," *Proceedings of the Geologists' Association*, vol. XVI, p. 449 and Marr, *Jubilee Volume of the Association* (1910), p. 624.

[3] Dakyns, J. R., "Notes on the Geology of the Lake District," *Geological Magazine* (1869), Decade I, vol. VI, pp. 56, 116.

The approximation to horizontality is still better shewn in the hills on either side of Ullswater. There the junction is almost horizontal, though itself thrown into shallow folds. The volcanic rocks and also the Skiddaw Slates are seen abutting against the plane of junction, and the outcrop of this plane on the uneven ground is extraordinarily sinuous, and in one case two masses of the volcanic rock above the plane are actually isolated from the main outcrop and entirely surrounded by the Skiddaw Slates. These isolated patches are represented on the one-inch Geological Survey Map shewing the lower reaches of Ullswater: they are on the north side of the lake at the western end of its lowest reach. We also find isolated patches of Skiddaw Slates surrounded by rocks of the Borrowdale Series, as near Scarf Gap Pass, Buttermere.

It is now generally admitted that the junction is one of faulting and not of unconformity, and it is mapped as such by the Geological Surveyors. The evidence to which we have referred above further shews that this junction is of the nature of a plane, sometimes removed about 30° from the horizontal, at other times much more nearly approaching it.

The question still remains whether this gently inclined fault is an overthrust accompanied by inversion.

No detailed statement has yet been made regarding the evidence on account of which the idea of inversion has been advocated, but many geologists have been struck by the general lithological resemblance between the volcanic rocks of the Borrowdale Series, and others occurring elsewhere which are of Precambrian age, especially those of Charnwood. The resemblance to the Charnian rocks is especially marked in the case of the flinty ashes and breccias of the Scawfell Ash and Breccia group. There are however lithological differences between these ashes and the Charnwood rocks. The former are calcareous and marked by an absence of visible quartz and detrital mica. It must be remembered moreover that these flinty ashes owe their present characters not only to their original composition, but also to their subsequent alteration, and rocks of very different ages might be altered in a similar way. Though these flinty ashes occur in the greatest abundance in the group just named, they are by no means absent elsewhere. Flinty ashes occur in thin bands associated with the lavas of the Falcon Crag group, by the roadside just south of the Rosthwaite alluvial flat in Borrowdale. They are quite similar in everything except extent of their development to those of the Scawfell group. These ashes are here folded in with the vesicular lavas in a remarkable manner. A similar association is seen on the left bank of Church Beck Coniston, close to the fall. Now these ashes are associated with lavas and other volcanic rocks which are as closely comparable with the Ordovician volcanic rocks of North Wales and Shropshire as are the flinty ashes with those of Charnwood. Consequently, it appears that on lithological grounds we may as justly claim an Ordovician age as a Precambrian one for the Borrowdale Series.

The abundance of garnets at any rate in all the lower divisions of the Borrowdale Series has been held by some to indicate great age. As garnets occur in lavas of various ages in other countries this goes for little as a test of age.

It may be remarked here that although garnets are so abundant in the rocks of the Borrowdale Series, they have not been detected in the Eycott group on the north side of the district. Their apparent absence there may possibly indicate that these rocks, notwithstanding resemblances to those of the Ullswater group, may yet turn out to be of different age, though I think that this is unlikely.

It has been hinted before that the junction between the Coniston Limestone and Borrowdale Series may be an unconformity or a fault. If it be an unconformity, it is explicable whether the Borrowdale rocks be newer or older than the Skiddaw Slates, though there are difficulties if the latter view be adopted.

It may be urged that the accumulation of the many thousands of feet of rocks of the Borrowdale Series must have taken a long time, and that the Llandeilo period was not of sufficient length for their formation. It has already been noticed that the fauna of the highest Skiddaw Slates suggests Llandeilo affinities, and that therefore perhaps only a portion of Llandeilo times was available for the formation of the volcanic rocks. On the other hand, we find to the east of the Eden Valley fossiliferous Caradocian rocks (the Roman Fell group) which are older than any Caradocian rocks of the Lake District, and in the latter therefore lavas and coarse ashes and agglomerates of the Borrowdale Series may have been accumulated while marine sediments were being deposited on an adjoining area, as long ago suggested by Goodchild. In that case the Borrowdale Series in the district may have been formed during the greater part if not the whole of Llandeilo times and also in early Caradocian times.

We may now consider the evidence which appears to me to militate against the hypothesis of inversion.

In the first place allusion must again be made to the fact (remarkable on the view of inversion, but quite according to expectation if the Borrowdale rocks are still in their original position), that we find volcanic rocks interstratified with the highest Skiddaw Slates and also with the strata of the Coniston Limestone. Not only is that the case, but those of the Skiddaw Slates are similar as regards composition and lithology with those of the oldest (Falcon Crag) group of the volcanic series, being andesitic, while similarly those of the Coniston Limestone like the rocks of the highest division of the Borrowdale Series are rhyolitic. This seems to be a remarkable coincidence if the volcanic rocks of the Skiddaw Slates and Coniston Limestone are not genetically connected with those of the Borrowdale Series.

It has I believe been suggested that the rhyolites at the top of the Borrowdale Series belong not to that series but to the Coniston Limestone, but this would only dispose of half of the difficulty, and furthermore the field evidence certainly seems to point to the rhyolites at the top of the Borrowdale Series really belonging to it.

Again, we find evidence in regions outside the Lake District of great volcanic activity in Llandeilo times, and we should not therefore be surprised to obtain similar evidence in the district itself. Also, as has been pointed out, the volcanic rocks of the district do present lithological resemblances to those of Ordovician age elsewhere.

Lastly, if we are dealing with inversion along the fault-plane between the Skiddaw Slates and the Borrowdale Series, and if the Coniston Limestone rests unconformably upon older rocks, it would appear that the thrusting must have taken place before the deposition of the Coniston Limestone, though after that of the formation of the Skiddaw Slates, for the Coniston Limestone is not only discordant to the strike of the Borrowdale rocks, passing on to older and older members of that series to the west of Coniston, but it actually rests upon the Skiddaw Slates themselves close to where they lie beneath rocks of the Borrowdale Series on the east side of the Duddon estuary in the neighbourhood of Dalton-in-Furness. Now some of the minor changes which have resulted from the movements of which the fault-plane at the base of the Borrowdale Series is one effect, are also noticeable in rocks of Silurian date, as will presently be described; hence it would appear that these movements were post-Silurian.

From what has been said it will be gathered that I regard the evidence in favour of inversion at the base of the Borrowdale Series as inconclusive, but after what we have learned of abnormalities in the succession in regions of greatly disturbed rocks elsewhere, we must be prepared for the possibility, that, notwithstanding apparent reasons to the contrary, inversion may have actually occurred here also.

Let us pass on to the second hypothesis, that the Borrowdale Series has lagged behind during a general onward movement. The general southerly dip of the strata and gently inclined fault-planes would indicate that such movement would be due to pressure exerted from the south. As already stated, if lag-planes occur, they must be subsidiary to a general overthrust, which need not necessarily be now visible, for it might be concealed beneath newer strata to the north of the Lake District. In this connexion, however, we must take account of the remarkable position of the Drygill Shales of Caradocian age. They are in apposition to the Eycott volcanic group on the north, and though they have intrusive rocks to the south, the Skiddaw Slates are found immediately south of these. The presence of the Drygill Shales might be due to an unconformity, but the lithological characters of the beds do not suggest this; the strata have the appearance of fairly deep water deposits. If there is no unconformity, the beds must have been brought into their present position by faulting. No exposure shews the junction between these beds and the underlying rock, so that the hade of the fault, if such exists, is unknown. It is possible that these Drygill beds actually occur below an overthrust fault, which has been here brought to a high level and exposed by erosion, owing to movements subsequent to the formation of the fault and perhaps connected with the injection of the intrusive rocks of Carrock Fell. It may be noted that these beds shew lithological characters very dissimilar to those of the Coniston Limestone further south, and this dissimilarity may be due to strata once far apart being brought nearer to one another by thrusting. The section in Fig. 13, which is purely diagrammatic, shews the relationship which would exist between thrust-plane, lag-faults, and the various rocks on this hypothesis.

I have elsewhere suggested that the higher Ordovician and the Silurian rocks on the east side of the Eden Valley, may lie below the overthrust fault, but it is only a suggestion and requires further work for verification or disproof.

Let us turn to the suggested lag-faults. It has been noted that in certain circumstances newer rocks may be brought over older by a thrust-fault, and this may have occurred along the junction between the Borrowdale rocks and the Skiddaw Slates. The question can only be settled by detailed mapping.

We may now regard the junction between the Coniston Limestone and the Borrowdale Series. It has already been stated that this has been claimed as an unconformity and also as a fault. I shall presently have more to say with regard to the material actually observed along the junction.

I have previously suggested that we are here dealing with a lag-fault, and supposed that it was more steeply inclined to the horizontal along the greater part of its outcrop, as indicated by the more regular line, as compared with the zigzag line of the fault at the base of the Borrowdale Series. Mr J. F. N. Green has however shewn that around the Duddon estuary, this junction, which he maintains to be one of unconformity, is at a low angle dipping "inward towards the estuary on each side with an inclination of less than 10°." The occurrence of the outlier of Coniston Limestone at Greystone House near Duddon Bridge on volcanic rocks, which as Green shews are not at the top of that series, is of course explicable on the hypothesis of fault or unconformity. .

The fault at the base of the Coniston Grits is also one which over a large part of the area appears to possess a considerable inclination. That a fault does occur here is shewn by the sudden cessation of the

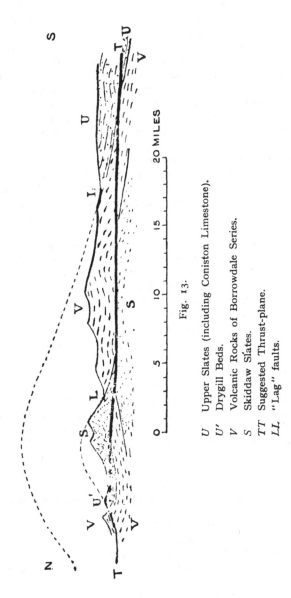

Fig. 13.

U   Upper Slates (including Coniston Limestone).
U′  Drygill Beds.
V   Volcanic Rocks of Borrowdale Series.
S   Skiddaw Slates.
TT  Suggested Thrust-plane.
LL  "Lag" faults.

tears against it, and also by the sudden diminution in the thickness of the upper part of the Coniston Flags on one side of each tear.

There is evidence of packing of the rocks so as to increase their apparent thickness, both in the Borrowdale Series, to the south of the fault which separates them from the Skiddaw Slates, and also in the Silurian rocks above the reputed fault at the base of the Coniston Limestone. The more rigid volcanic rocks have been packed by fracture, giving rise to pseudo-stromatism on a large scale, as well seen in the rocks of the Scawfell Ash and Breccia group in the central part of the district. The less rigid Silurian rocks have been largely packed by folding: this is well shewn in the case of the Bannisdale Slates. The packing of the rocks has not only taken place parallel to the strike but also at right angles to it; hence we find the rocks are thickest in the centre, thinning out to the east and west. This is shewn in the case of the rocks of the Borrowdale Series by the small thickness of those rocks along the district near Shap on the east, and around the Duddon estuary on the west.

Again, among the Silurian rocks we find the great basin of the Kirkby Moor Flags along the tract which is apparently that of greatest packing; this is a true basin and not merely a syncline, and its longer axis is actually in a general north and south direction, that is approximately at right angles to the general folding of the Lower Palaeozoic rocks of the district.

We come now to the third hypothesis, that the faults are not lag-faults, but overthrusts along a plane more nearly horizontal than the original planes of junction of the rocks. From what has been said above, the hypothesis of lag-faults is only tentative: it may prove that this third hypothesis is after all correct. As already stated, the question must await settlement until more detailed mapping has been carried out. If either of the two last hypotheses be correct, the rocks of the Borrowdale Series must be newer than those of the Skiddaw Slates. The crucial question is whether the first hypothesis or one of the two latter be correct, and as I have already said, it appears to me that the evidence as a whole militates against the first hypothesis.

Let us now consider more fully some of the minor changes which were brought about as the result of the movements which we are considering.

The localities and directions of the larger folds have already been considered. There are several points of interest in connexion with the minor folds. The Skiddaw Slates are frequently intensely folded. Both normal folds and overfolds are frequent, and the septa of the latter are often affected by a thrust-plane. The volcanic rocks, as already stated, are more frequently faulted than folded. Certain minor folds which are subsidiary to the faulting will be considered presently. The intensity of the pressure in these rocks in some cases is illustrated by the extraordinary contortion of a dyke cutting through banded ashes on Long Crag north of the Three Shire Stone at Wrynose.

The beds of the Coniston Limestone Series are often contorted on a small scale. Contortions also occur in the Stockdale Shales. I have figured an overfold in the Skelgill beds of Browgill, and much isoclinal folding is noticeable among the *Dimorphograptus* shales of these beds to the south-west and south of Coniston.

The Coniston Flags have been affected by cleavage rather than by folding, but the Coniston Grits often shew well-marked minor folds. This is still more noticeable in the Bannisdale Slates. Situated between two fairly rigid rock groups—the Coniston Grits and Kirkby Moor Flags —they have undergone much contortion, often of the nature of over-folding, with frequent accompaniment of thrusting along the septa. Much complex folding of these beds may be seen in numerous small quarries in the neighbourhood of Bowness and Windermere Villages, but it is even more marked along portions of the eastern side of the district. The Kirkby Moor Flags, like the Coniston Grits, frequently shew minor folds of a simple nature.

It has been observed that the major faults are also accompanied by minor ones and of these we shall say more.

Normal faults are frequent, and require no further notice, but concerning the faults having fissures approaching the horizontal and the tear-faults a few more remarks are advisable.

The approximately horizontal faults no doubt occur of all degrees of magnitude and may even be seen in microscopic sections, as for instance in the fine banded ashes of the Scawfell Ash and Breccia group. The frequent occurrence of lenticular outcrops of rock in the Borrowdale Series, separated from one another by turf-clad ground, may be due to a network of curved faults of this type, for where the faults are certainly absent on a small scale, the rocks often have outcrops which are continuous for long distances, as for instance in the case of those of the Falcon Crag group.

The Skelgill beds of the Lake District are everywhere affected by a nearly horizontal fault which usually separates the beds of the *Dimorpho-graptus* zone from those of higher horizons, but it sometimes occurs above the top of the *Dimorphograptus* zone. Some years since a quartz-vein was seen along a minor fault of this type in Skelgill, the upper surface being polished to so great an extent by the friction of overlying rocks that it had a lustre almost mirror-like. It has since been covered by a landslip.

A very important feature in connexion with these faults is that where a yielding stratum has been situated between more rigid masses, horizontal movement has taken place at the top and base of the yielding stratum, and it has become violently puckered, and even brecciated. The puckering can be simulated by sliding the hands over a piece of tissue-paper placed between them. In these cases the contorted beds are underlain and overlain by rocks with perfectly regular bedding. They are frequent among the Borrowdale Series, especially in the banded ashes of the Scawfell group, where the structure is well brought out owing to the fine lamination of the yielding beds. Beautiful examples of this structure occur near Walney Scar. I figure a striking

Fig. 14.  Contorted and brecciated Volcanic Ash
S.E. of Sty Head Tarn.

example shewing both contortion and brecciation on a large block above Sty Head Tarn, which forms a semi-detached portion of a band traceable for a long distance (Fig. 14).

The same structure is developed in the Bannisdale Slates, where laminated mudstones lying between grits have been affected in the way described. Good examples have been found in Windermere Village, and may now be seen in the blocks of a high wall by the road which passes below the railway bridge at the signal-box, just east of Windermere station. A similar structure is seen at Mearness near Greenodd.

The importance of this observation lies in the fact that there is every gradation between the horizontal faults which have produced this structure on a small scale and the larger faults which are to some extent accompanied by such structures on a much larger scale, including the fault at the base of the Borrowdale Series. But as the structure was produced after the deposition of the Bannisdale Slates, the faults and folds of various degrees of importance are thereby dated as of a period subsequent to the deposition of the Bannisdale Slates, that is to post-Silurian times, and not to a period prior to the deposition of the Coniston Limestone.

The tear-faults also require further notice. The manner in which they die out against the nearly horizontal faults has already been mentioned. Another proof that they indicate horizontal movement along vertical planes is furnished by localities where they cut anticlines or synclines. It is well known that where folds are cut by normal dip faults, the outcrops of a bed on the two sides of the axis of the fold come between those of the same bed on the other side of the fault-plane, whereas when the folds are affected by tears, the outcrops on the two sides of the fissure either alternate or are completely separated. Also that in the former case the outcrops in the two sides of the fissure are at different distances apart, while in the latter the distances are equal. Displacement of the latter type shewing the existence of a tear-fault is seen in the case of a syncline in the Kirkby Moor Flags north of Whiteside Pike, between Long Sleddale and Bannisdale.

Another indication of horizontal movement is furnished by the *vertical* Armboth dyke which is laterally shifted by faults upon Armboth Fell. This is not only of interest as proving horizontal movement along the fault, but it fixes the date of the dyke as anterior to that of the movements which caused the faults.

A very interesting occurrence between two parallel tear-faults is noticeable in the great cliff just above the lower bridge at Skelgill, and is illustrated in the accompanying section (Fig. 15). It will be noticed that the beds between the *Atrypa flexuosa* zone and that of *Monograptus argenteus* are much thinner between the two tear-faults than on either side, so that a thickness of a few feet which is represented outside the tears has been plucked away by horizontal movement between the tears, and accordingly the side of each fault which shews a downthrow in the case of the *argenteus*-zone shews an upthrow in the case of the *flexuosa*-zone.

We may now make further allusion to the fault-breccias. In many cases it is difficult to distinguish these from the explosion-breccias of the Borrowdale rocks, but in others the brecciation is clearly the effect of subsequent movements. The breccias are associated with faults whose fissures approximate to the vertical and also to the horizontal: in the latter case, fracture of the rock has usually taken place to a much greater extent than in the former, and there is indication in places of ultimate complete crushing of the fragments, giving rise to mylonitic belts.

In the case of the vertical tear-faults, it has been noted that brecciation often affects wide belts, and that these belts, being bands of weakness, have frequently been stained red by percolation and

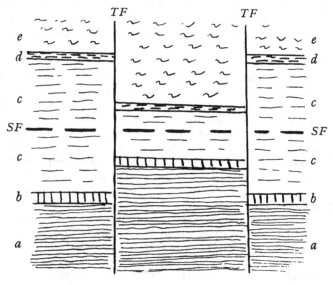

Fig. 15.

| | | | |
|---|---|---|---|
| *e* | *Phacops elegans* Zone. | *a* | Ashgill Shales. |
| *d* | *Monograptus argenteus* Zone. | *SF* | Strike Fault. |
| *c* | Beds between *d* and *b*. | *TF* | "Tear" Faults. |
| *b* | *Atrypa flexuosa* Zone. | | |

deposition of iron compounds. The same is also the case with the nearly horizontal fault-breccias. It has already been mentioned that the great purple breccia of Borrowdale (which is represented on William Smith's Geological Map of Cumberland) is very probably a fault-breccia, and a similar origin is probable for many of the purple breccias which occur at various horizons in the Borrowdale Series.

In some of these fragmental rocks movement has gone on to an extent sufficient to round off the original angular fragments, thus giving rise to fault-conglomerates. A good example of a conglomerate of this character is shewn at the base of Wolf Crag, at the northern end of the Helvellyn range. This conglomerate contains pebbles of garnetiferous lava. As some of the lavas in the upper part of the Falcon Crag series are garnetiferous, the fault-conglomerate may have been formed between the Falcon Crag and Ullswater groups, the former being here faulted out. At least three bands of conglomerate occur in the rocks of the Falcon Crag series on the north side of Brown Dodd near Lowdore. They may be contemporaneous deposits, but it is possible that they are fault-conglomerates.

Similar conglomerates are found near Butterwick, on the east side of the district, near the junction with the Skiddaw Slates. They are also found in many other places in the heart of the Borrowdale Series.

A curious pseudo-concretionary structure is sometimes seen, and is well shewn to the west of Picthall Ground on the east side of the Lower Duddon Valley; it appears to be due to incipient brecciation. Again in places a pseudo-perlitic structure seems to be due to brecciation on a small scale. A rock of this type is well shewn on Crowberry Haws, Coniston.

A word may be said about the "bird's-eye" slates so well seen in the Kentmere quarries. The "bird's-eyes" are probably concretions, but a rock in Caudale quarry on John Bell's Banner near Kirkstone Pass seems to indicate that the concretionary structure is developed around breccia-fragments. These bird's-eye slates are widely distributed. They are found in many quarries along the Kentmere Slate band. They also occur near Po House, Millom, and on the north side of the district, in the banded ashes below Gillercombe.

The possible occurrence of a fault-breccia at the base of the Coniston Limestone Series has already been mentioned, but it remains to notice certain rocks which occur on the eastern side of the Duddon estuary.

A junction between the Skiddaw Slates and the Borrowdale Series was seen in a quarry on the mineral railway at the foot of Crag Wood, west of High Haulme near Dalton-in-Furness. The two rocks were welded together and both crushed. A still more interesting section is seen in Hole Gill, in the village of Ireleth, below the bridge on the Kirkby-Ireleth road. The junction is seen a foot or two above the stream for many yards, on either bank, and finally disappears at the base of a waterfall under the bridge. The rocks of the two series are much crushed and a mixed crush-breccia often marks the junction. It contains "eyes" of a volcanic breccia. A similar section is seen further north at Bank House, Souterthwaite.

I would venture to suggest that the "brown clay crammed with andesite cores" which Green describes in a cutting he made in the rocks of Greenscoe Crags, lying between the Coniston Limestone and the volcanic series, may also be fault-stuff. He objects to the existence of a fault here, because the junction is folded in the same way as a bedding-plane. As a similar folding occurs in the fault at the base of the Borrowdale Series around Ullswater, this occurrence hardly militates against the existence of a fault near Dalton.

Little need be added to what has already been said about the cleavage. A brief chapter on the cleavage by Clifton Ward will be found in the Geological Survey Memoir on the Geology of the Northern Part of the English Lake District, and it is accompanied by a map shewing the positions of the anticlinal and synclinal axes of the bedding-planes and cleavage-planes. At times the anticline of one coincides with the syncline of the other: at others it does not.

Otley notices that in Borrowdale the cleavage-planes dip to the north, and in Langdale to the south, which would indicate that the syncline of the Scawfell group of fells may coincide with an anticline of the cleavage-planes.

Ward mentions a highly cleaved felsite dyke on Kirk Fell. This should be noted, as here we get one test of age of an igneous intrusion, which has so far not been greatly utilised. It is obvious that this dyke was injected before the cleavage and therefore probably belongs to the earlier set of intrusions.

We will now further consider the mineralogical changes. In the first place it is probable that the conversion of many rocks into a flinty state, so well exemplified in the case of the rocks of the Scawfell Ash and Breccia group in the Scawfell hills, is accompanied by some mineralogical change, though further microscopic examination of these rocks is required to settle this. This flinty structure was no doubt developed during the general period which we are considering, for these rocks have escaped cleavage. The agent which produced the flinty structure is unknown. It may have been due to igneous intrusion, but the fact already noticed, that such flinty structure occurs on a small scale elsewhere, suggests that it was more probably due to hydrothermal action. Other mineralogical changes are no doubt the result of pressure, probably accompanied by action of heated water. Among these is the development of white mica, already noticed as having occurred in the rocks of all these divisions. The production of mica from crushed felspar in lavas of the Borrowdale Series was referred to earlier in the chapter. This change is especially well marked in highly cleaved rocks in which the cleavage-planes have been subsequently bent into sharp chevron-bends, of which several may occur in the space of an inch. Such rocks are veritable gnarled schists. A beautiful example occurs behind Watendlath, on the right bank of the gully which is just above the hamlet. Similar rocks are seen north of the lower end of Haweswater. These rocks approximate to mylonites. A true mylonite, where the rock seems to have been crushed to powder, and a certain amount of mineralisation has taken place, is found in the slate-band in a quarry near Low in the Duddon Valley. Similar rocks must occur elsewhere and should be looked for.

The changes in the garnets to which allusion has been made may be seen in many places, as on the fells behind Rosthwaite, and those near Wrynose Pass. The garnet becomes crushed and partially converted into white mica. As the result of further change this passes into chlorite. A central core of unaltered garnet may often be observed, but ultimately the garnet disappears entirely, and the chlorite remains as dark green pseudomorphs after garnet.

The study of mineralogical changes in the various rocks, which were produced during the movements we have described, has not been carried out extensively by examination of microscopic rock-sections.

Much information concerning the microscopic characters of some of these rocks is given by W. Maynard Hutchings[1].

# CHAPTER XII

## THE CHANGES AT THE END OF LOWER PALAEOZOIC
### TIMES (*continued*)

We have still two questions to consider. Firstly the characters of the intrusive igneous rocks of the Devonian period, and secondly the nature of the erosion which took place in Devonian times.

The rocks of the newer suite which are enumerated by Harker[2] are the granites of Skiddaw, Eskdale and Wastdale, and Shap; various sills and dykes of quartz-porphyry; mica-lamprophyre intrusions; the Carrock Fell intrusions; and dolerite and andesite dykes. The general age of some can be determined with certainty, others with probability—while that of others again is extremely doubtful. It will be convenient however to consider in one chapter all the rocks of the newer suite which are intrusive in the Lower Palaeozoic rocks of Lakeland.

---

[1] Hutchings, W. M., "Petrological Notes on some Lake District Rocks," *Geological Magazine* (1891), Decade III, vol. VIII, p. 536, and "Notes on the Ash-Slates and other Rocks of the Lake District," *ibid.* (1892), Decade III, vol. IX, pp. 154, 218.

[2] Harker, A., *Proceedings of the Yorkshire Geological and Polytechnic Society*, vol. XIV, p. 491.

We will consider the principal masses in turn, and at the same time treat of those dykes and sills which give evidence of genetic connexion with the large masses. When describing each mass, it will be convenient also to give an account of the character of the alteration which it has produced in the surrounding rocks.

Four large masses, each probably of a laccolithic nature, are intrusive into the Lower Palaeozoic rocks. These are the granites of Shap, Skiddaw, Eskdale and Wastdale, and the rocks of varied composition which form the intrusive mass of Carrock Fell.

### The Shap Granite.

We may begin with the Shap granite, of which the age can be definitely fixed. The granite and its metamorphic effects have been described by Dr Harker and myself[1]. It is situated at the eastern end of the district on the Shap Fells. It occurs in the form of an irregular oval having a longer diameter from east to west of a little under two miles, and a shorter north and south diameter of somewhat over a mile.

The rock as seen is everywhere in contact with rocks of the Borrowdale Series, though it may be in contact with the Coniston Limestone at the extreme south where the rock is covered with detritus and vegetation. (See map, Fig. 16.)

The rock consists of two felspars, quartz and black mica, forming a matrix of moderately coarse texture. In this matrix are embedded numerous flesh-coloured crystals of felspar, one or two inches in length. The matrix is generally grey, but in places becomes red. The two coloured varieties are usually spoken of as the "light" and "dark" granite. There is every reason to suppose that the red rock is merely a modification of the normal grey granite due to subsequent change. The rock is not very acid, and has a silica percentage of about 68·6.

Apart from the colour variety, there is a considerable

---

[1] Harker, A. and Marr, J. E., *Quarterly Journal of the Geological Society* (1891), vol. XLVII, p. 266 and *ibid.* (1893), vol. XLIX, p. 359.

uniformity of character in the rock over the whole exposure, with the exception of some marginal modifications.

Everywhere the rock includes some dark patches of varying size: such are known to Scotch quarrymen as "heathen." The "heathen" of the Shap granite are in some cases inclusions of earlier rock caught up in the granite, but others are clots of igneous rock caught up in the normal granite. The latter often contain the large pink crystals of felspar characteristic of the granite itself, though in the case of the "heathen" they have undergone certain changes along the margin.

Radiating out from the granite, often to a distance of several miles, are a series of dykes, and in addition there are frequent small sills. These sills and dykes extend as far as the Sedbergh and Pennine districts, but they become much more abundant as the granite is approached. They vary much in character but belong to two main groups of rock, the quartz-felsites and the lamprophyres or mica-traps. The former are more and the latter less acid than the granite. That they are genetically connected with the granite is shewn among other things by the existence of the characteristic crystals of pink felspar in the felsites and mica-traps alike. There are also other mineralogical resemblances between these dykes and the granite; the lamprophyres especially are very closely connected with the dark clots with pink felspars which are found in the granite.

The evidence points to the derivation of granite, quartz-felsite and lamprophyre from one igneous magma or molten reservoir, probably having the general composition of the granite. This molten rock became partly separated into liquid material of two kinds, which were injected into the overlying rocks to form felsitic and lamprophyric dykes and sills, while a part of the magma was forced upwards in a molten condition without undergoing separation, and thereby retained its original composition. This formed the present mass of Shap granite, which apparently is of the nature of a laccolith formed of subparallel sheets meeting at the centre, and thus presenting in cross-section a rude resemblance to a cedar, whence the type is termed "cedar-tree laccolith."

After the granite was consolidated heated vapour and water caused the deposition of various minerals, including fluorspar, malachite, iron-pyrites, copper-pyrites and molybdenite.

Let us now consider the evidence bearing upon the age of the granite. It has been seen that the visible junction is always in contact with rocks of the Borrowdale Series: this proves merely that the granite was consolidated after the formation of those rocks. But the dykes radiating from the granite cut through rocks of all ages up to and including Upper Ludlow rocks, and, as we shall subsequently see, rocks of various ages, including Ludlow strata, are metamorphosed as the result of the intrusion of the granite. The granite is therefore post-Ludlow, in other words post-Silurian in age.

Fortunately we have also means of ascertaining the latest possible date of its formation. In Blea Beck, near Shap Wells, is a conglomeratic deposit, once supposed to be of Old Red Sandstone age, but now usually regarded as the base of the Carboniferous System. Amongst the fragments in this conglomerate are abundant rolled crystals of the typical pink felspar of the Shap granite. The granite had therefore been intruded, and the rocks above it had undergone erosion to a degree sufficient to lay bare the granite and allow of its erosion when this conglomerate was laid down. Thus the granite is pre-Carboniferous and post-Silurian, in other words it was intruded in Devonian times.

In a strictly petrographical sense the Shap "granite" is not a true granite, being separated from the true granites by its porphyritic character. Further noticeable characters of the rock are the abundance of plagioclase felspar in addition to the orthoclase, and the nature of the quartz, which gives evidence of consolidation before the orthoclase. The rock belongs to the adamellites.

In addition to the minerals already enumerated we may notice the presence of apatite, zircon, frequent magnetite, and more especially sphene, the latter being always present and often abundant.

Some of the marginal modifications of the granite are of interest. We described a peculiar rock found as a loose block below the footpath north of Wasdale Beck. I have since found this rock *in situ* between the high road and the quarry railway at the south-east end of the granite. The rock has a marked parallel structure which gives it a laminated

appearance suggestive of sedimentation, and the phenocrysts of orthoclase are rather rare. Microscopic examination of the rock shews that
it is truly igneous and the banding is a fluxion-structure. A notable
feature of this rock is the abundance of andalusite therein.

Occasionally we find marginal rocks of coarser texture than the
normal granite. One of these occurs on the quarry railway near a stile
west of the "Lodge." It is characterised by peculiar "blade-like"
mica-crystals; they are long narrow blades, sometimes an inch in length,
with irregular terminations. A similar rock is found on the hillside
about 350 yards N.W. of Wasdale Head Farm.

On the west side of Sherry Gill the junction between granite and
rocks of the Borrowdale Series is marked by a pegmatitic vein, a rather
coarse-grained aggregate of felspar and quartz, in which much of the
former has been replaced by the latter mineral.

Regarding the basic inclusions we may notice that they have a much
lower silica percentage than that of the normal granite, viz. 56·95 as
opposed to 69·6 of the granite. They are specially characterised by the
great abundance of sphene and dark mica. It is indeed the latter
mineral which gives them their prevailing dark colour. There is also
a predominance of triclinic over monoclinic felspar among the smaller
crystals. The phenocrysts of orthoclase are rounded by corrosion of
the margins, and have an outer white corrosion-ring in which the
orthoclase has been converted into plagioclase and quartz.

The characteristic features of these inclusions are reproduced in the
lamprophyric dykes and sills, and indeed such differences as are observable between the two are due to the weathering which has subsequently
affected the intrusions, but which the inclusions have escaped.

The genetic relationship of the lamprophyres and felsites with the
granites is further borne out by the fact that the orthoclase phenocrysts
of the former are most abundant as one approaches the granite and
become rarer as we pass away from it. They are nevertheless found
sparingly so far away as Swindale Beck, near Knock, on the east side
of the Eden Valley.

Attention may be called to the characters of some of the more
important dykes and sills. Two ordinary quartz-porphyries occur near
Wasdale Head Farm, which contain sphene, but otherwise call for no
further remark. A large quartz-porphyry dyke four or five hundred
yards south of Wasdale Old Bridge contains porphyritic crystals of
white monoclinic felspar sometimes two inches in length. They are in
the form of partly interpenetrating Carlsbad twins and have the
characters of sanidine.

An interesting claret-coloured felsite with quartz-crystals and
tolerably large felspar phenocrysts is seen at the head of the plantation
in the Shap Wells grounds.

In the stream south of Wasdale Beck several sills of felsite and mica-trap occur in the neighbourhood of Stakeley Folds, about two-thirds of a mile from the granite. They contain felspar phenocrysts, which in the case of the lamprophyres are modified like those of the basic inclusions of the granite.

Further down this stream, in the plantation at the Gill Farm, is a composite dyke, part of which is felsitic, the rest lamprophyric. Of dykes further afield we may notice one on Potter Fell, which has micas of blade-like habit like those of the marginal modification of the granite alluded to above[1].

Let us now turn to the metamorphic effects produced by the granite upon the surrounding rocks. These are of interest on account of the great distance to which the metamorphism extends. The production of new minerals is confined to distances of from 1200 to 1300 yards from the granite contact, that is to a distance about equal to the mean semi-diameter of the granite mass itself as exposed at the surface. The rocks are slightly affected to a still greater distance as evidenced by hardening. Another interesting feature is the great variety of rocks which have been altered. They consist of intermediate and basic andesitic lavas, rhyolitic lavas, and the associated tuffs and agglomerates of these rocks; all these belong to the Borrowdale Series. In the Coniston Limestone are pure and impure limestones, calcareous shales and calcareous ashes. Among the Silurian rocks we find shales, calcareous shales and grits. There is also an altered fault-breccia and a metamorphosed mineral vein or rather plexus of veins.

The nature of the alteration of the volcanic rocks is largely affected by the changes that had occurred in them previous to the intrusion of the granite. It is clear that they had been affected by ordinary weathering, which caused the kaolinisation of the felspar and the conversion of the more basic minerals into chlorite; this is accompanied by the refilling of the vesicles of the lavas with various minerals, as calcite, chlorite and quartz.

The lowest volcanic rocks around the granite are basic

---

[1] For further details concerning the lamprophyres see Harker, A., "The Lamprophyres of the North of England," *Geological Magazine* (1892), Decade III, vol. IX, p. 199.

andesites, and these are succeeded by intermediate andesites. As regards the general character of the metamorphism, there is little difference in the case of the two varieties. A progressive increase in the amount of metamorphism is observable as one approaches the granite. The contents of the vesicles first undergo mineralogical change. Nearer the granite the change begins to be developed in the mass of the rock, and increases until the whole has become recrystallised. Near the junction the vesicles become indistinct, their contents being partly merged in the general recrystallisation of the rock.

The main changes in these rocks are the production of a purple-brown mica, of felspar and of quartz in the bulk of the rock, and of quartz, hornblende and red garnet in the vesicles.

Above the andesites we meet with a rhyolitic series—the Shap Rhyolitic Group. This consists of lava-flows and ashes. In the case of the lava-flows the metamorphic changes are often slight, as these rocks were generally less weathered prior to metamorphism than were the andesites. Some of the rhyolitic ashes have, however, undergone interesting changes.

In some of the ashes at a distance from the granite spots are developed which are of a crystalline character, and are markedly contrasted with the rock around them.

Nearer the granite the changes are more marked, and recall to some extent those which have occurred in the andesitic rocks, except that there is a smaller proportion of the purple-brown mica, and, on the other hand, there is in places a development of the mineral kyanite. This mineral is found for instance in the rocks on either side of the high road near the south-east margin of the granite. The rocks at this place were banded ashes, and as the result of the metamorphism they have been converted into true schists.

The Coniston Limestone consists of a great variety of deposits, and accordingly the products of alteration are very numerous, but, as one would expect, the dominant feature is the production of various lime-silicates.

Both the Stile End beds and the Applethwaite beds of the series are found in a highly metamorphosed condition just west of Wasdale Head Farm. The former may be found in two small sikes, but it is difficult to detect the exposures. The latter is exposed in a stream by the cart-track west of the farm and by the side of the track itself. The two divisions are separated by the Yarlside rhyolites, the alteration of which does not differ in any important respect from that of the Shap rhyolites. The lower limestone, which was here impure, is converted into idocrase, lime-garnets and other lime silicates. The upper limestone, also impure, is largely changed into a greyish porcellanous-looking rock. The most abundant mineral is lime-augite, but lime-felspar is far from rare. Other lime-containing minerals which are found in both Lower and Upper Limestone are tremolite and wollastonite.

No highly metamorphosed examples of the Stockdale Shales have been detected. If not faulted out they are concealed by superficial accumulations near the granite.

The Brathay Flags, on approaching the granite, become spotted rocks. Ultimately the whole rock becomes recrystallised, the principal minerals developed being purple-brown mica, quartz and felspar, so that the changes in general resemble those of the ashy members of the volcanic rocks.

The Lower Coldwell Grits and Coniston Grits have been converted into quartzites, and some new minerals are developed from various impurities. The calcareous Middle Coldwell Beds exhibit a type of metamorphism generally similar to that of the Coniston Limestone, with formation of lime-silicates.

A fault-breccia is seen on Packhorse Hill between the Lower and Middle Coldwell Beds, and is metamorphosed. More interesting is the metamorphism of the plexus of mineral veins seen in a quarry in the rhyolitic group north of Blea Beck and west of the high road. The veins before metamorphism contained quartz, calcite, chlorite, iron and copper pyrites, and galena. The metamorphism of the bulk of the rock is similar to that of the vesicles of the andesitic lavas, with development of quartz, hornblende, recrystallised calcite and garnet. The garnets are very numerous and of large size, often exceeding an inch in diameter. The effect of the metamorphism on the metalliferous contents has not been studied in detail.

We may add some further remarks about the character of the meta-morphism in the various rock-divisions.

In the first place it may be noted that the development of some minerals has taken place along cleavage-planes, and thus we obtain clear proof that the cleavage, which as already remarked succeeded the general folding and faulting of the Lower Palaeozoic rocks of the district, was itself anterior in date to the intrusion of the granite.

The chloritoid mineral of the volcanic rocks which undergoes change into the purple-brown mica appears to be delessite. The conversion of this mineral in the vesicles into hornblende instead of mica is no doubt due to the presence of calcite in the vesicles. This hornblende usually occurs in the form of dark green acicular crystals. The reconstituted felspars of the volcanic rocks are, like those of many of the schistose rocks of a regionally metamorphosed area, of the " water-clear " character. This " water-clear " felspar has the property of resisting kaolinisation as the result of weathering action. The garnets of the vesicles of the lavas, like those of the metalliferous vein, differ from the light-coloured lime-garnets found in the altered Coniston Limestone. They are of a deep-brown colour proper to the lime-iron-alumina garnets.

In the rhyolitic ashes close to the junction with the granite Mr Hutchings detected sillimanite and andalusite[1].

Details concerning the very numerous minerals of the Coniston Limestone group may be sought in our papers on the Shap rocks. One point may be noticed, namely, the abundance of tremolite and other magnesian minerals in the altered rocks, which suggest that the lime-stone may have been partly dolomitised before its metamorphism. Dolomite crystals do, in fact, occur in some of the unaltered limestones of Blea Beck near the mineral spring in the grounds of Shap Wells Hotel.

In addition to the minerals mentioned as occurring in the altered Brathay Flags, Hutchings records hornblende, garnet and anatase[2]. The last-named mineral appears to be formed at the expense of the rutile-needles which occur abundantly in the unaltered rock. In places there is a development of white mica in veins, and it also penetrates into the rock around the veins. The nature of the spots in the spotted rock has not been satisfactorily cleared up. Some appear to consist almost wholly of white mica. A large proportion are probably anda-lusite. In this connexion it may be remarked that no large crystals of an alumina-silicate mineral such as are common in argillaceous rocks around an igneous mass have been found in these Brathay Flags. The material of some of the spots is " beyond any attempt at identification."

---

[1] Hutchings, W. M., "Note on a Contact-Rock from Shap," *Geological Magazine* (1895), Decade IV, vol. II, p. 314.

[2] Hutchings, W. M., "Notes on the Altered Coniston Flags at Shap" *Geological Magazine* (1891), Decade III, vol. VIII, p. 459.

Taking the metamorphosed rocks as a whole, it may be noted that the aureole of metamorphism cannot be divided into a series of zones as is often done in the case of aureoles around an igneous mass. This is probably due to the great diversity of composition of the rocks which strike against the granite mass. These rocks have each undergone alteration according to their capacity to become changed, and accordingly the lines differentiating the various zones would occur at different distances from the granite contact in the case of rocks of different composition.

Study of the metamorphic rocks around the granite tends to shew "that thermometamorphism is not in general accompanied by any change in the chemical composition of the rocks affected" with the exception of partial loss of water and expulsion under certain conditions of carbonic acid: further it would appear "that no transference of material has taken place within the mass of the rocks except between closely adjacent points....If this be...the general law, it follows that the mineral produced by complete thermal metamorphism at any point of a rock depends upon the chemical composition of the rock-mass within a certain small distance around that point."

### The Eskdale-Wastdale Granite[1].

This granite extends from the foot of Wastwater in a southerly direction to Bootle Fell, having a length of twelve miles with a maximum breadth of four miles. A smaller outcrop of the same granite is found near Wastdale Head. It is intrusive into the rocks of the Borrowdale Series, and on the west Triassic rocks lie against it, but unfortunately no junction is seen between the Trias and the granite. At its northern end it comes against the Ennerdale granophyre, and here again no contact of the two rocks has been observed. The normal granite is of a distinctly acid character, containing about 75 per cent. of silica, and it becomes still more acid at the margins. The rock is white to grey and pink to red. It consists of quartz, orthoclase and plagioclase felspars (and an intergrowth of the two), and white and black micas, the former usually predominating. At Waberthwaite a variety occurs with an exceptional amount of black mica.

[1] The granite and surrounding rocks are described by Prof. A. R. Dwerryhouse, "On some Intrusive Rocks in the Neighbourhood of Eskdale (Cumberland)," *Quarterly Journal of the Geological Society* (1909), vol. LXV, p. 55.

A large number of dykes occur around the granite, but of these Professor Dwerryhouse considers that only one is connected with it. It runs "from Great Bank, near Eskdale Green (at which locality it occurs as a vein in the granite itself), by way of the south-eastern slope of the Screes to Wasdale Hall at the head of Wastwater. Thence it passes across Lingmell Gill and the north-western flank of Lingmell to the granite exposure of Wasdale Head, which it appears to penetrate. On the opposite side of the granite, it can be traced from the neighbourhood of Burnthwaite up the side of Eskdale Fell, and across the summit of Kirk Fell to Bayscar Slack, at the foot of Kirkfell Crags." It is a reddish rock of the quartz-porphyry group with "well-marked porphyritic crystals of felspar and quartz, with a smaller quantity of mica, in a micro-granitic ground mass," and agrees petrologically with the granite itself. Prof. Dwerryhouse gives evidence of the laccolithic character of the main intrusion, for fortunately portions of the original covering of the granite, consisting of rocks of the Borrowdale Series, still remain on its summit as outliers in two places, namely Blae Tarn Hill and Great Barrow, west of Eskdale.

The age of the granite is doubtful. It is newer than the Borrowdale rocks and almost certainly older than those of Triassic age, but there is no definite evidence to shew whether it belongs to the older or newer suite of volcanic igneous rocks. The evidence, so far as it goes, is in favour of its newer age. As Dwerryhouse remarks, "it seems probable that this intrusion belongs to the Devonian Period, as does the neighbouring granite of Shap, which, with the exception of its large phenocrysts of orthoclase, is not dissimilar to some of the varieties of the Eskdale granite."

The acid margin of the granite is of interest, and appears "to furnish another illustration of the general law, that those substances which crystallise first from the average magma—an acid one in this case—tend to concentrate on the cooler parts." Dwerryhouse gives it as his opinion that "the occurrence of the more acid rock on the periphery and the more basic in the interior, is due to the magma as a whole having been more acid than some eutectic, probably that of quartz and orthoclase, and the consequent primary separation of quartz until the eutectic proportions were reached, at all events locally."

It is interesting to note that Dwerryhouse only mentions one dyke as definitely connected genetically with the granite. Walker mentioned some dykes in the Scawfell tract as possibly having such a connexion. I always felt that their petrographical characters shewed closer relationship with the igneous rocks of the earlier suite, and am glad to find that Dwerryhouse also considers them to be unconnected with the Eskdale granite.

The metamorphism around the granite has not been worked out in detail. Dwerryhouse in his paper gives some notes on the characters of the metamorphic rocks. They are very similar to those of the Borrowdale Series altered by the Shap granite, but though the exposed area of the Eskdale granite is far greater than that of Shap, the metamorphism does not extend so far outward from the granite margin in the case of the former rock as in that of the latter. Harker and I suggested as a reason for the exceptional metamorphism around the Shap granite that it occupies the position of a pipe, possibly connected with the surface, up which igneous rock flowed for some time before the final consolidation of the present granite mass. A good example of the metamorphism is seen in Oliver Gill, where rocks of the Ullswater group are in contact with the granite. They are rocks coloured by the purple-brown mica with green hornblende developed in the vesicles of the lavas, just as at Shap. The similarity of the type of metamorphism in the case of the Shap and Eskdale granites is to some extent confirmatory of the view, which is supported by petrographical resemblances between the two rocks, that they are both of Devonian age.

### The Skiddaw Granite[1].

The Skiddaw granite occurs in three separate but neighbouring outcrops, in a general north-south line. The most southerly patch is seen in Sinen Gill, a tributary of the Glenderaterra on the west side of Saddleback. The second and

---

[1] See Harker, A., "Carrock Fell: A Study in the Variation of Igneous Rock-Masses, Part II," *Quarterly Journal of the Geological Society* (1895). vol. LI, p. 139.

largest is in the Caldew Valley, and the third occurs lower
down that valley, and is also seen in its tributary, Grainsgill.
The nature of the exposures indicates that here also we are
dealing with a mass of laccolithic character, for there is evidence
that the junction between the granite and the overlying rock
approaches the horizontal. The normal rock "is essentially
a biotite-granite consisting of orthoclase, oligoclase, quartz,
and brown mica." There are often scattered flakes of muscovite. The rock is medium-grained and of a light grey
colour. A specimen in White Gill has a silica percentage of
75·223.

In the two southern exposures the granite is normal, but in
the third and most northerly it becomes more acid, and the
acidity increases on passing northward until the rock is
eventually a greisen, that is a felsparless rock composed of
quartz and white mica.

A specimen of the modified granite from the Caldew 300 yards above
the junction with Grainsgill Beck gave a silica percentage of 77·26 while
a sample of the greisen near the foot of Brandy Gill has one of 78·13.
Harker regards the difference between the greisen and the normal
granite as due to differentiation, owing to mechanical force operating
on the granite-magma when crystallisation had proceeded to a certain
stage. The difference is increased as the result of pneumatolytic action.

The greisen and the rocks in its vicinity are penetrated by
numerous mineral veins containing many rare minerals, which
have rendered the Caldbeck Fells classic ground to the mineralo-
gist. Some at any rate of these veins may be due to the
action of vapours and heated waters marking the final stage
of igneous activity; the earlier stages being marked by the
intrusion of the igneous rock[1].

It has been seen that certain mica-traps have a genetic
relationship with the Shap granite. Two occurrences of mica-
trap are noted in the northern part of the district and have been
described by Clifton Ward in the Survey Memoir so frequently

[1] For information concerning these and other minerals of the District the
reader may consult "Contributions towards a List of Minerals Occurring in
Cumberland and Westmorland," by J. G. Goodchild, printed in the *Transactions of the Cumberland Association for the Advancement of Literature and
Science*, Parts VII, VIII and IX.

quoted (p. 33). One occurs on Sale Fell, west of the lower end of Bassenthwaite Lake. It has a pink crystalline felspathic base containing both orthoclase and plagioclase, the former being more abundant. In this is much dark-greenish mica, probably biotite, and a little quartz.

Several mica-trap dykes are seen on Skiddaw Dodd. They consist essentially of felspar and mica with some quartz. It is probable that these lamprophyric rocks are genetically connected with the Skiddaw granite, from which they are not far distant.

The age of the Skiddaw granite is uncertain. As in the case of that of Shap metamorphic minerals have been produced along the cleavage-planes of the Skiddaw Slates, and accordingly the granite was injected after the occurrence of the earth-movements which produced that cleavage. It almost certainly belongs to the newer suite of igneous rocks, and is probably of Devonian age.

The general nature of the metamorphism produced on the Skiddaw Slates into which the granite is intruded has long been known and is described by Clifton Ward in "The Geology of the Northern Part of the English Lake District" (pp. 9–12) He considered that there were three zones around the granite, the outer being one of chiastolite-slate, the middle of spotted slate, in which the rocks are thoroughly recrystallised and contain in addition numerous crystalline spots of aggregates of an alumina silicate, and the inner of mica-schist. Subsequently the researches of R. H. Rastall shewed that this simple arrangement does not altogether hold good, though substantially correct with regard to the various metamorphic rocks of the Glenderaterra Valley[1]. He finds here (what was seen to occur in the neighbourhood of the Shap granite) that rocks of original different lithological characters strike against the area in which the granite is found, and that these rocks are altered in varying degrees, so that concentric zones of metamorphism cannot be continuously traced. The accompanying map (Fig. 17) is a reduction of Rastall's map of the metamorphic area.

[1] Rastall, R. H., "The Skiddaw Granite and its Metamorphism," *Quarterly Journal of the Geological Society* (1910), vol. LXVI, p. 116.

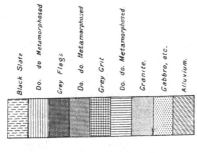

Black Slate

Do. do. Metamorphosed

Grey Flags

Do. do. Metamorphosed

Grey Grit

Do. do. Metamorphosed.

Granite.

Gabbro, etc.

Alluvium.

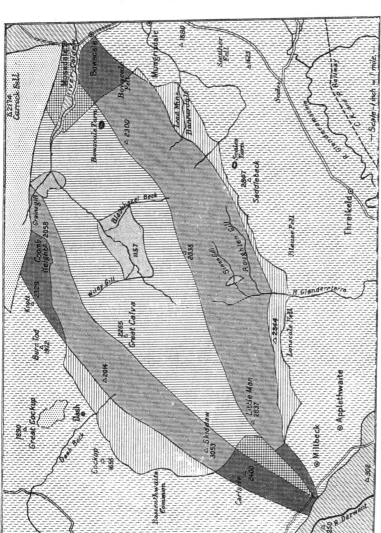

Fig. 17. Map shewing metamorphic rocks around the Skiddaw Granite.

The rocks consist of a central band of gritty deposits developed in the Caldew Valley. North and south of these are grey flaggy beds, and outside of these soft black slates.

The beds are certainly repeated by a fold which on the whole appears to be an anticline, though the numerous minor complications render this doubtful, and Rastall suggests that the granite itself may have been injected along the core of this supposed anticline.

The Black Slates never come within the influence of the area of extreme metamorphism along their outcrops. In these slates as one passes inward from the non-metamorphic area the first sign of alteration is often the occurrence of spots, giving rise to an inconstant outer spotted zone. The nature of the spots is doubtful, but they appear to be embryo crystals, in some cases, of chiastolite. As one passes to tracts where metamorphic changes are greater chiastolite crystals become abundant and the well-known chiastolite-slate is developed. The crystals are sometimes $1\frac{1}{2}$ inches long. The comparatively feeble nature of the metamorphism of these chiastolite-slates is proved by the occasional discovery of fossils in them.

The Grey Flags at the outer edge of the aureole of metamorphism undergo somewhat similar changes to those seen in the Black Slates. Spots begin to develop at the outer edge, and further in chiastolite is found. Nearer to the central portion of the aureole the mineral cordierite, first detected by Harker in these rocks, becomes an important constituent, that and biotite being the dominant minerals. Close to the granite of Sinen Gill the Grey Flags are converted into the so-called mica-schists. In these also the dominant minerals are those of the cordierite rock, but in immediate contact with the granite at Sinen Gill garnet and staurolite are also developed.

The Grey Grits, as might be suspected from their siliceous character, have on the whole undergone less alteration than the other rocks, many of the altered rocks resembling impure quartzites. Next to quartz the most abundant mineral is brown mica, which is largely replaced by chlorite in the neighbourhood of the granite. This rock may have been affected by vapours emanating from the granite. Cordierite is also found in the altered rocks of this group, and in the immediate vicinity of the granite, garnet occurs.

The evidence afforded by the minerals developed in the metamorphic aureole points to "the maintenance of a moderate temperature for a long period of time, thus allowing of a very complete recrystallisation for a long distance from the margin." Accordingly there appears to be an entire absence of such minerals as kyanite and sillimanite which are formed at high temperatures.

"The phenomena here displayed may be very briefly summed up as an example of a moderate degree of thermal metamorphism due to the intrusion of a great mass of granite, at a comparatively low temperature, into a series of rocks of somewhat variable composition, but on the whole rich in alumina. The most important minerals produced are cordierite, andalusite and its modification chiastolite, biotite and muscovite; while garnet and staurolite are only found close to the granite. Owing to the great variation in lithological character, it has not been found practicable to divide the aureole into definite zones, but the alteration is gradual and progressive from without inwards."

### The Carrock Fell Intrusion[1].

The remarkable and varied series of igneous rocks which form a large part of Carrock Fell and the adjoining fell region are of particular interest as illustrating the differentiation of a rock-magma. The mass has a length of about $3\frac{1}{2}$ miles along its longer axis from east to west, and is over a mile wide at its widest part. Harker gives reasons for supposing that we are here dealing with a mass of laccolithic character and it may be suggested as a possibility that it was intruded along a fault-plane.

The accompanying map by Harker (Fig. 18) shews the distribution of the igneous rocks in this tract. The granite and greisen belong to the Skiddaw granite already described, and Harker gives reasons for regarding the diabase as later in date than the gabbro and granophyre. It is these two rocks with which we are concerned.

---

[1] The Carrock Fell igneous rocks are described by A. Harker, "Carrock Fell: A Study in the Variation of Igneous Rock-Masses, Part I," *Quarterly Journal of the Geological Society* (1894), vol. L, p. 311; "Part II," *ibid.* (1895), vol. LI, p. 125.

There is evidence of two periods of differentiation of a magma. The gabbro and granophyre shew genetic relationships which indicate that they are the products of differentiation of a single magma. After this differentiation, the gabbro was first intruded into its present position, and during intrusion itself underwent further differentiation. Subsequently, and while the gabbro was still hot though solid, the granophyre was injected, and produced change in that part of the gabbro which was in contact with it. The later date of the intrusion of the granophyre is clearly proved by the fact that it sends veins into the gabbro. The rocks were intruded between the Skiddaw Slates, which lie to the south, and the Eycott volcanic rocks and Dry Gill Shales, which are now found on the north-east and north. Furthermore a large mass of Eycott rocks is now found on the summit of the gabbro, indicating the laccolithic character of the intrusion.

There is evidence that the rocks surrounding the intrusion have been affected by the post-Silurian crustal movements prior to the formation of the igneous mass, and furthermore that the rocks of the mass itself have not been affected by these movements. This would place the mass as one of those belonging to the newer suite of igneous rocks. Many of these were formed in Devonian times, and that may be the case with the Carrock Fell rocks, but they may even be later in age.

The essential minerals of the gabbro are a triclinic felspar and a monoclinic pyroxene, while quartz and iron ores are marked constituents of the more acid and more basic types. The rock is of a greyish colour, of medium or moderately coarse grain, exceptionally becoming very coarse. It is divided into a series of belts which appear to have vertical or nearly vertical junctions. The central belt is the most acid, and from the development of quartz is a quartz-gabbro. On either side of this is what may be termed the normal rock in which quartz is absent. On the outer margin are belts with felspar reduced to a minimum, and a great increase in the amount of titaniferous iron ore. The change is quite gradual and orderly, and is undoubtedly due to differentiation of a magma which probably had originally a composition approximating to that of the normal gabbro of the mass.

The change when passing from centre to margin is shewn by the diminution of the silica percentages, and the rise of the specific gravity. The highest silica percentage found in the quartz-gabbro is 59·46, while the lowest in the iron-ore bearing gabbro of the margin is 32·53. Between these two extremes we have intermediate amounts when passing from centre to margin.

The variations in the specific gravity are shewn by Harker as the result of three traverses taken across the rock. I give one: it will be seen that the specific gravity decreases steadily until the central zone (C) is passed:

$$3·265$$
$$\overline{\phantom{xxxx}}$$
$$2·850^1$$
$$2·822 \ (C)$$
$$2·890$$
$$2·939$$
$$3·110$$

This increasing basicity of the gabbro from centre to margin " is explained by a concentration of the basic elements in the cooler portion of the magma during the progress of crystallisation."

The granophyre is an augite-granophyre. It shews scattered crystals of black augite and whitish plagioclase felspar in a fine-textured grey, cream-coloured or reddish ground mass. The latter shews a granophyric structure due to the intergrowth of minerals. It consists chiefly of quartz, and orthoclase and plagioclase felspar. The granophyric structure shews a considerable variety of micropegmatitic and spherulitic characters.

A specimen of the rock of average composition gave a silica percentage of 71·60, whilst the most acid rock analysed furnished one of 77·38.

Along the junction between the gabbro and granophyre is a belt of rock of very considerable interest. It is usually very

---

[1] The line represents a place where the rocks were concealed.

coarse-grained and consists of felspar, augite, hornblende, iron-ore, and other minerals. Some of the crystals have a length of two inches. Harker shews that this rock was probably due to the incorporation of masses of yet heated basic gabbro in the molten granophyre.

It has been seen that the gabbro shews orderly variations due to differentiation, and that the granophyre is more uniform throughout. Nevertheless each of the rocks shews local variations, the more important of which may be pointed out.

The gabbro often shews a banded structure, where rocks of different textures alternate in thin layers, simulating stratification. The origin of this structure is doubtful, and Harker considers that it cannot be regarded as a flow-structure.

In addition to the mass of volcanic rock on the summit of the gabbro several masses are embedded in it. Where the gabbro is in contact with these rocks, a certain amount of absorption of the material of the volcanic rocks by the gabbro has occurred, with alteration in the composition of the latter, which is marked by a development of brown mica.

The variations in the characters of the granophyre are unimportant save that along the southern margin, where its composition has been changed, as already described, by an absorption of some of the material of the already consolidated gabbro.

A certain amount of metamorphism has been produced on the surrounding rocks by both gabbro and granophyre, but it has not been studied in detail. The Skiddaw Slates are much altered on the southern margin, but much of this alteration is undoubtedly due to the adjacent Skiddaw granite. On the north side the junction is so much obscured by superficial accumulations that little can be made out.

It has been seen that a change of the nature of "inverse metamorphism" has been produced in the gabbro by the absorption of some of the volcanic material. The gabbro has also produced direct metamorphism on these volcanic rocks when included in or floated upon the summit of the gabbro. The metamorphism, unlike that observed in the case of the Shap granite, is not very great. Details of the changes which have taken place will be found in Harker's paper.

### Diabases, Dolerites and Andesites.

Intrusive rocks of this character are of various ages, and little is known about them. Some of them certainly belong to the newer suite of intrusive rocks, and many of them may be of

Devonian age.   I have already referred to the diabase which
is newer than the Carrock Fell gabbro, but is probably of the
same general geological age.   In this connexion one may
notice the record by Groom of a tachylyte traversing the
gabbro of Carrock Fell[1], and similar rocks are also seen in the
granophyre, just below the summit of the fell.   Many of these
more basic rocks are no doubt of date later than Devonian
times, though, when found among the Lower Palaeozoic rocks,
their date is a matter of uncertainty.   Some dykes are seen
cutting the conglomerate which lies at the base of the Carboniferous rocks near the foot of Ullswater, and these are probably
post-Devonian.   Harker remarks that a few dykes have been
observed "which resemble known Tertiary dykes, and are
perhaps to be referred to that period.   They are apparently
less basic on the whole than the older dolerites, and they are
in a fresher condition."

These rocks should properly have been considered in a
later chapter, but, as they are few in number and of no great
interest, it has been thought best to consider all the rocks
referred to the newer suite of intrusive rocks in one and the
same chapter.

### *Erosion at the end of Lower Palaeozoic Times.*

A study of the geological map of the Lake District at once
brings out the fact that between the rocks of the Lower and
Upper Palaeozoic groups there is a great unconformity.   This
is best seen along the line of junction between the two groups
in the tract of country extending from the foot of Ullswater
to Tebay.   Here the base of the nearly horizontal strata of
Carboniferous age traverses the various upturned members
of the Lower Palaeozoic divisions from the Skiddaw Slates of
Ullswater to the Kirkby Moor Flags near Tebay.

It is perfectly clear therefore that the earth movements
which affected the older strata and caused them to be folded
into an anticline with its axis through the Skiddaw hills, and
also caused the host of minor folds, faults and other changes

---

[1] Groom, T. T., "On a Tachylyte associated with the Gabbro of Carrock
Fell," *Quarterly Journal of the Geological Society* (1889), vol. XLV, p. 298.

which we have considered, took place before the deposition of the Carboniferous rocks, so that the upper part of the anticline was cleared away by erosion, causing the oldest strata now visible along the line of its axis to be exposed at the surface.

There is little doubt that as the result of the movement the district was converted into a land area, and that much of the erosion was of a subaerial character. The greatest amount of erosion on the whole no doubt took place where the oldest rocks are now exposed, and if we had a definite estimate of the thickness of the rock removed, we could calculate the amount of maximum erosion. The thickness of rocks given in the table in the introductory chapter adds up to about 30,000 feet or more, but it does not follow that all these strata attained their maximum thickness where the Carboniferous rocks are now seen resting upon them. Indeed we know that what is probably the thickest group, the Borrowdale Series, is comparatively thin at the eastern margin of the district. Nevertheless we shall ultimately see reason for supposing that the Carboniferous rocks once extended over the whole district and in its central parts the united thickness of the whole probably approaches, if it does not exceed, the figure given. This estimate approximates to that given by Clifton Ward[1]. He argues for the removal by erosion of 20,000 to 25,000 feet of strata and says: "to this amount of removed material we must add a considerable thickness of Skiddaw Slates, themselves cut from the dome top." Ward's estimate of the thickness of the Borrowdale Volcanic Series is considerably less than that given later by Sir A. Geikie, and recent work certainly shews that Ward's estimate is under the mark. It is perfectly clear then that along the line of axis of the uplift very many thousand feet of strata were removed by erosion, and that 30,000 feet is probably not an over-estimate. That this erosion affected the highest Silurian rocks and preceded the deposition of some of the oldest Carboniferous rocks indicates that it took place in Devonian times, and this accounts for the absence of sediments of that age in the district.

[1] Ward, J. C., "On the Physical History of the English Lake District," *Geological Magazine* (1879), Decade II, vol. VI, pp. 50, 110.

The Carboniferous rocks now dip away from the district:
if we restored them to their original condition before they were
affected by subsequent movements, we should find them
resting upon a comparatively, though by no means absolutely,
flat surface, which would separate them from the Lower
Palaeozoic rocks beneath.   This surface was a plain of denudation, and it remains to be seen whether it was a peneplain or
a plain of marine denudation, in other words whether the erosion
was entirely subaerial, or whether the final levelling was produced by marine action.   Concerning this there is little evidence.
The earliest of the Upper Palaeozoic rocks shew some indication
of having been formed as terrestrial accumulations, and in
places they certainly seem to have accumulated in valleys
such as might have been excavated by stream-action.   This
suggests a peneplain, but much more work on this subject is
required before the matter can be regarded as settled.

All we can now say is that the marine Silurian period in
Lakeland was separated from the marine Carboniferous period
by a Devonian continental period when the Lower Palaeozoic
marine strata had been elevated into land which was subjected
to the ordinary agencies of erosion.


# CHAPTER XIII

## THE CARBONIFEROUS ROCKS

The Carboniferous rocks do not enter into the Lake District
proper, but there are reasons why they should be briefly
regarded here.   In the first place, their nature and trend has
a bearing upon the events which occurred in the district after
their formation: again, as the development of these rocks is
of considerable interest, anyone who desires to get a full
knowledge of the geology of the region will probably visit
some of the sections shewing the sequence of Carboniferous
strata; lastly, though the rocks are not in the district, they
are of it, for some very beautiful scenery can be seen in the

estuarine tracts at the south of the district, where these rocks are developed, and also from the heights of the Carboniferous hills to the west and south-west of Kendal.

The three great divisions of the Carboniferous system— Carboniferous Limestone, Millstone Grit and Coal Measures— are developed in the vicinity of Lakeland, for the Cumbrian coal-field lies at no great distance from it, on its north-western side. It is however the Lower Carboniferous rocks which more directly concern us, and to these we may confine our attention. There are two marked lithological developments of these rocks, of which the lower is largely sandy and con-glomeratic, and the upper calcareous. They may be spoken of as

Carboniferous or Mountain Limestone,
Basement Conglomerate.

### (a)  *The Basement Conglomerate.*

The Basement Conglomerate is chiefly developed along the eastern margin, though there are traces of it to the north. The Carboniferous Limestone extends all round the district except where the newer Triassic strata come against the Lower Palaeozoic rocks on the west, between Whitehaven and Millom.

The rocks of the basement series are of a sandy nature, usually coloured deep red by iron oxide. Some of the beds are pure sandstones, while others contain pebbles of various sizes, in places up to three feet in diameter. These pebbles are usually sub-angular or even angular. They consist in some cases of rocks of local origin; thus at Hutton, north of Ullswater, are numerous pebbles of Skiddaw Slates and of rocks belonging to the Borrowdale Series, and the presence of crystals from the Shap granite in these beds near Shap Wells has already been noted. In addition to these local pebbles are many others which have been carried for long distances. Everywhere we find pebbles of Silurian grit, and at Hutton are pebbles of limestone, which Clifton Ward states to be probably Coniston Limestone. As no Silurian rocks are developed near Ullswater it is obvious that these pebbles have been borne for some distance, and as Ward suggests, probably from the south. The beds often shew variations which appear to be due to successive floods; coarse material alternates with "patches of current-bedded, fine-grained gravelly material, representing the action of the feebler stream which continued after the passage of the flood[1]."

The idea that these deposits were accumulated in old valleys was long ago suggested by Prof. Phillips and Mr Godwin-Austen[2].

Two views have been expressed as to the age of these deposits—the one that they are of Old Red Sandstone age, and the other that they form the base of the Carboniferous strata of the district. In the absence of fossils it is impossible to settle the question, but considering the great changes that took place before they were deposited, and after the accumulation of the highest Silurian strata of the district, considering also that the succeeding Carboniferous rocks are closely associated with the basement conglomerates, we may treat these conglomerates as marking essentially the beginning of the Carboniferous period, and the recent work of Prof. Garwood shews that the succeeding Carboniferous rocks lie conformably on the conglomerates. The rocks vary much in thickness, their maximum on Mell Fell being 800 or 900 feet.

The exact mode of transport of the pebbles of this polygenetic conglomerate is not yet settled. Ice-action has been invoked to account for it, but the general opinion now is that there is no evidence of such action. Oldham[3], after examining the deposits at the foot of Ullswater, draws the conclusion "that the conglomerate is a torrential deposit, formed on dry land, near the foot of a range of hills, in a generally dry climate, varied by seasonal or periodical bursts of rain. The red colour of the fine-grained material suggests tropical or sub-tropical conditions, as the formation of red soils is at the present day so much more common

---

[1] Oldham, R. D., "The Basal (Carboniferous) Conglomerate of Ullswater and its mode of Origin," *Report of the British Association*, 1900, p. 764.

[2] See references in the Survey Memoir, "Geology of the Northern Part of the English Lake District," p. 76. Details concerning these Basal Conglomerates are given in this Memoir, also in the Memoirs on the Geology of the Country between Appleby, Ullswater and Haweswater and of the Country around Kendal, Sedbergh, Bowness and Tebay.

[3] *loc. cit.*

in tropical than in temperate regions that it may almost be regarded as a characteristic of a hot climate." It is not necessary that all the deposits of this series should have been formed under exactly the same conditions, and parts of them may have been laid down in the waters of shallow seas. Dr Strahan calls attention to the unevenness of the platform of Silurian rocks on which the conglomerates rest near Shap Wells, and remarks that "the conglomerate is seen to be placed over and round such rock-masses as may be seen on a modern sea-shore[1]."

In this connexion the occurrence of a marine band recorded by Prof. Garwood (in a paper to be referred to presently) should be noticed. It is found in Pinskey Gill in Ravenstonedale, and contains brachiopods and other marine fossils. Garwood believes that it is overlain by the red conglomerates of the Basement Conglomerate Series.

The occurrence of intrusive basic rocks in the conglomerate near the foot of Ullswater has already been noted. They occur on and near the roadside east of Little Mell Fell. A description will be found in the Survey Memoir[2].

### (b)  The Carboniferous Limestone.

We have already stated that the evidence indicates that the basal conglomerates form the true base of the Carboniferous rocks of the area. The detailed mapping by Garwood along the line from Ullswater to Ravenstonedale indicates that this conglomerate is succeeded conformably by the next sub-division of the Carboniferous sequence, which here forms the lowest part of the Carboniferous Limestone Series. The conglomerates and limestones have been considered separately on account of their great lithological differences.

The Carboniferous limestone consists typically of beds of white or grey-white limestone, with a certain amount of mechanical sediments, which are more marked towards its base and summit. This limestone is no doubt mainly of organic origin, though in many cases the organisms have been obliterated. Various animals contribute to its formation, including crinoids and brachiopods, but Garwood has recently shewn that many of the limestones are essentially composed of calcareous algae.

[1] "Geology of the Country around Kendal, Sedbergh, Bowness and Tebay," p. 23.

[2] "Geology of the Country between Appleby, Ullswater and Haweswater," p. 61.

The important bands of limestone are known locally by geographical titles, which will be found on the maps and memoirs of the Geological Survey, but Garwood in a most exhaustive paper has divided these deposits zonally, and correlated the various zones of different parts of the area, and also those of the area as a whole with those of other areas[1].

Two types of the limestone may be spoken of as the northern and southern types. The northern type is that which is developed around Lakeland proper, but the southern type approaches our area in the Kirkby Lonsdale district, near Carnforth, also to the south and south-east of Dalton-in-Furness. I shall have occasion ultimately to refer to the significance of these two types.

Garwood describes four main zones in the Carboniferous limestone; they are, in descending order, as follows:

> *Dibunophyllum* zone,
> *Productus corrugato-hemisphericus* zone,
> *Michelinia grandis* zone.
> *Athyris glabristria* zone.

These are further subdivided into ten sub-zones.

These zones prove that the pre-Carboniferous land-surface in the north-west of England "does not appear to have been submerged until some time after the Carboniferous sea had invaded the South-Western Province," as is shewn by the entire absence in the north of the lowest zones of the Bristol district. But "the lowest deposits met with in the North-Western Province occur in close proximity to the present margin of the Lake District, namely between Pooley Bridge and Shap Wells, at the foot of the Howgill Fells in Ravenstonedale, at Meathop in the Arnside district, and near Elliscales and Marton in the Furness district." The significance of this statement will be seen when we discuss the nature of later movements.

The most satisfactory tracts around the district for studying the succession are in the strip of country between Shap and Ravenstonedale,

---

[1] Garwood, E. J., "The Lower Carboniferous succession in the North-West of England," *Quarterly Journal of the Geological Society* (1912), vol. LXVIII, p. 449.

the Fells west of Kendal, and the borders of the Kent estuary about Arnside. Those who desire to make acquaintance with the detailed succession of the strata are referred for information to Prof. Garwood's paper[1], but a few remarks may be added upon two points.

Firstly we may note "the important transgression of an arenaceous type of deposit across several of the faunal horizons that occur in the Shap and Ravenstonedale districts, traces of which can also be found in the Pennine district on the east and the Kendal district on the west This sandstone episode, which begins at the base of the *Michelinia* Zone at Shap, does not make its appearance at Ravenstonedale until the middle of the Lower *Productus-corrugato-hemisphericus* sub-zone; it also persists up to a higher horizon in the former district than in the latter."

More important, for a reason which will be considered later, is the existence of the northern and southern types to which reference was made above.

The contrast between two types of rock on either side of the Craven fault has long been known, and R. H. Tiddeman described a remarkable series of knoll-like masses of limestone containing a peculiar fauna to the south of that fault[2]. Similar rocks occur nearer to our district. Professor Hughes calls attention to resemblances between the Lower Carboniferous rocks of the Kirkby Lonsdale area and those to the south of the Craven fault[3], and as has been observed Garwood shews that these rocks, like those between Carnforth and Lancaster, and others south and south-east of Dalton-in-Furness, are of the southern type. The beds between Carnforth and Lancaster shew knoll-like structures similar to those which are found further to the east.

Along the Craven fault the two types are found in contact, and, as will be seen subsequently, the same thing occurs along a fault at the southern margin of Lakeland.

---

[1] Garwood, *loc cit.*, p. 553.

[2] Tiddeman, R. H., "Report of the International Geological Congress, 4th Session, 1888 [1891]," p. 319. See also "Report of the British Association for 1889," p. 600.

[3] *Geological Survey Memoir*, "Geology of the Neighbourhood of Kirkby Lonsdale and Kendal," p. 22.

# CHAPTER XIV

## POST-CARBONIFEROUS CHANGES

The details of the development of the Permian and Triassic rocks do not concern us, for the rocks are not exposed in the district, but as study of movements which have affected the district necessitates some knowledge of these rocks, we shall say a few words about them. They are well developed in the Eden Valley to the north-east of the district, and extend from thence over the Cumbrian plain on the north and to the sea at Maryport. They set in again, as already stated, south of Whitehaven and extend thence to near Millom. Another mass occupies the country around Barrow-in-Furness, and there is a small patch around Cark-in-Cartmel. A more distant patch of importance to us lies south of the Craven fault at Westhouse near Ingleton.

The rocks consist essentially of red-coloured sandstones, with thin shales and limestones in places, the latter being confined to the rocks of the Permian System. The whole of the Permo-Triassic rocks give indications of having been formed under continental conditions during the prevalence of an arid climate. These continental conditions shew that a set of movements took place in post-Carboniferous times, and converted the old Carboniferous sea into land, just as at an earlier date the Lower Palaeozoic sea was converted into land in Devonian times. Though the general result was the same in each case, namely conversion of sea into land, the detailed effects of the movement were somewhat different so far as our district was concerned. It was seen that folding largely occurred in the case of the earlier movements, though it was accompanied by faulting, and that the former may even have been subsidiary to the latter. In the case of the post-Carboniferous movements which we are now considering, folding, except on a very small scale, was absent, and the result of the movements was essentially the production of a series of faults.

The movements were of the nature giving rise to plateaux rather than to mountain-chains.

That the movements we are now speaking of largely occurred in Permo-Triassic times is shewn by the fact that many of the faults bring Permo-Triassic rocks against those of earlier date. This merely proves that the faults were later in date than the deposition of some of these rocks. But Professor Kendall has brought forward evidence to shew that the movement along some of the fault-planes was actually proceeding in Permo-Triassic times[1]. He proves this in the case of the Pennine fault on the east side of the Eden Valley, and also in that of the Craven fault of the West Riding of Yorkshire.

There is evidence that these movements affected the rocks around the Lake District.

The fault-system of this date has never been worked out in detail. The more important faults are shewn in Fig. 19 and Fig. 20 is a section along the line $XY$ of the plan. I have elsewhere suggested that the Craven fault (strictly speaking the S. Craven fault, for there is more than one fault) is a thrust-plane, and that the comparatively undisturbed rocks to the north of the fault have been pushed southwards over the highly disturbed rocks to the south. There is evidence of the extension of this fault in a westerly direction into the district to the south of Lakeland. To the north of this fault-system are a series of north-and-south faults, which appear to be of the nature of tear-faults; along these horizontal slickensiding may frequently be observed. The Craven fault as generally regarded ends in a westerly direction against a pair of north-and-south faults of which the more westerly appears to be the newer, though no doubt they are of the same general age, and may be regarded together. One is the Dent fault, while to the west of this is the Barbon fault. These faults are traceable far northward, to the neighbourhood of Sedbergh. Some miles west, a similar north-and-south fault extends from near Kendal to the neighbourhood of Over Kellet, east of Carnforth, and there it ends against a fault striking north-west and south-east, which is traceable from Kellet to the foot of Warton Crag, where it brings the Millstone Grit on the south against the Carboniferous Limestone of the Crag; to the west of this it abuts against an apparent tear-fault, running north and south. It is probably shifted southward by

---

[1] Kendall, P. F., "The Brockrams of the Vale of Eden," *Geological Magazine* (1902), Decade IV, vol. IX, p. 510, and "The Geology of the Districts around Settle and Harrogate," *Proceedings of the Geologists' Association* (1911), vol. XXII, p. 27.

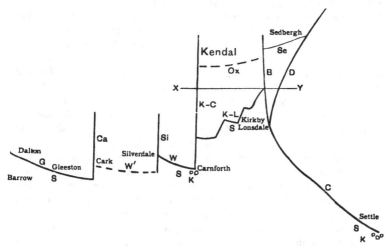

Fig. 19.

| | | |
|---|---|---|
| *C* | Craven Fault | |
| *Se* | Sedbergh Fault | |
| *Ox* | Oxenholme Fault? | |
| *K–L* | Kirkby Lonsdale Faults | Thrust-faults? |
| *W* | Warton Crag Fault | |
| *W'* | Westerly continuation of same? | |
| *G* | Gleaston Fault | |
| *D* | Dent Fault | |
| *B* | Barbon Fault | |
| *K–C* | Kendal-Carnforth Fault | Tear-faults? |
| *Si* | Silverdale Fault | |
| *Ca* | Cartmel Fault | |
| *S* | Southern type of Carboniferous Limestone. | |
| *K* | Knolls. | |

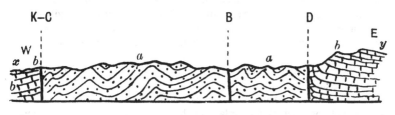

Fig. 20.

*b*  Carboniferous Limestone.      *D*  Dent Fault.
*a*  Lower Palaeozoic Rocks.       *B*  Barbon Fault.
          *K–C*  Kendal-Carnforth Fault.

this and other tears, to appear again south of Dalton-in-Furness, between that town and Gleaston, and is continued in a westerly direction separating the Permian rocks of the Barrow area from the Carboniferous rocks to the north. The phenomena along this fault-system are similar to those along the Craven fault-system. The Carboniferous rocks are highly folded, so that anticlines of limestone come between synclines of Millstone Grit, and in the anticlines the knoll-structure is developed. Furthermore the patch of Permian rock to the south of the fault at Barrow recalls that to the south of the Craven fault at Westhouse. Similar phenomena occur in the Isle of Man, so that this fault-system may actually extend through the south part of that island, and possibly even to Rush in Ireland where the knoll-phenomena are reproduced.

To return to the supposed tear-system: the Kendal-Carnforth fault has older rocks on its eastern side against newer on the west along the northern part of its course; west of the village of Holme there is a nodal point of no apparent vertical displacement where rocks of the same age abut against one another, while further south newer rocks on the east occur against older on the west. This is explicable upon the hypothesis of horizontal movement, otherwise we must suppose that there was a vertical see-saw movement with the nodal point as its fulcrum.

Now the rocks in the Kirkby Lonsdale district are of the southern type, though bordered to the east and west by those of the northern type. It would seem therefore that between the Dent-Barbon fault on the one hand and the Kendal-Kellet fault on the other, thrusting had not taken place in this latitude to the same extent as it did to the east and west. It looks as though two masses were driven southward along the chief tear-faults and that the anticline of Silurian rocks between these two areas where the southward movement occurred had escaped fracture on a large scale, though minor fractures are present. This escape may be compensated for by southerly movement in this tract at a different latitude, and we get evidence of the possibility of this further northward, where a general east-and-west fault has brought down the Old Red Conglomerate against the Silurian rocks of the Howgill Fells, while along the same line further westward two masses of low Carboniferous Limestone strata occur as outliers to the east of Oxenholme Station.

A large number of faults further north are undoubtedly connected with these movements.

I have entered into this matter at some length, as being connected with the changes which took place in the Lake District, and as shewing, what has no doubt been gathered from remarks made on other problems, that although much has already been done, a great deal of work yet remains to be accomplished in connexion with the Geology of the Lake District.

It has been argued at various times that the Carboniferous rocks once extended over the Lake District and that the district did not exist as an island against which the Carboniferous rocks were deposited. This seems to be borne out by the radial dip of these rocks, which is sufficient to carry them over the highest hills of the district, and confirmatory evidence is supplied by Garwood's conclusion already mentioned, that the lowest beds occur in close proximity to the present margin of the district. That some of these Carboniferous rocks were afterwards removed is shewn by the presence of Permo-Triassic rocks on the Lower Palaeozoic rocks, between White-haven and Millom. The dips of these Permo-Triassic rocks however are sufficient to carry them over the highest hill-tops, and Goodchild has concluded, I think justifiably, that they also once extended over the district. It remains now for us to consider when and how the Lower Palaeozoic rocks of Lakeland became exposed upon the surface.

After the accumulation of the Triassic rocks, the higher Secondary and the Tertiary strata were deposited in other areas. Long periods of time elapsed during the formation of these rocks, which with one exception are absent from the vicinity of the district, and we must briefly regard what happened in Lakeland when these rocks were being accumulated.

We have reason to suppose that at the end of Triassic times the area existed as a land-tract, probably of the plateau type, but diversified by block-hills and intervening depressions as the result of faulting, though these were probably to some extent smoothed down by erosion of the uplands and accumulation in the lowlands. At the end of Triassic times widespread submergence occurred, and that this affected the area at any rate contiguous to Lakeland is shewn by the presence of marine Liassic strata near Carlisle.

These Liassic rocks occur in the centre of the Eden Valley geological trough, which is an irregular syncline between the Lake District and Pennine anticlines. The occurrence of Lias beds in this syncline shews that at any rate some of the earth-movements of this area are post-Liassic. I have argued elsewhere that the later Mesozoic rocks and even some of the early Tertiary rocks may have been formed over what is now the Lake District[1].

We have evidence of two periods of considerable folding after Triassic times, which have affected the British rocks. One of these was in mid-Cretaceous and the other in mid-Tertiary times. If the movement which (with subsequent erosion) caused the appearance of the Palaeozoic rocks on the surface dates from one of these periods it is probably referable to the later one. The evidence however is by no means conclusive. It may be that this movement actually occurred in Triassic times, though I regard it as later.

There is a tendency to regard elevated tracts composed of ancient rocks as land-surfaces of ancient date. But we know that hard rocks of late geological age can be eroded into hill and valley tracts comparable with those occurring among ancient rocks, as for instance in the islands of Skye and Mull.

If the Lakeland region had been continuous land since Triassic times, and unaffected by any subsequent uplift, one cannot resist the conclusion that the whole area would long since have been base-levelled and converted into a peneplain.

The actual age of the uplift is however a subsidiary affair as compared with its nature.

So long ago as 1848 W. Hopkins, in a paper "On the Elevation and Denudation of the District of the Lakes of Cumberland and Westmorland," discussed the connexion between the geological structure of the district and the trend of its valleys[2]. He conceives the junction-surface of the Mountain Limestone and other formations as having been continued over the central portion of the district, and assigns the production of the dome to uplift subsequent to the deposition of the Triassic rocks, and indeed suggests later movement in Tertiary times. He argues that the radial drainage of the district is dependent upon the formation of the dome, and though he regards radial fault-lines as playing a more important

---

[1] See Presidential Address to the Geological Society, *Quarterly Journal of the Geological Society* (1906), vol. LXII, p. lxxxv.

[2] Hopkins, W., *Quarterly Journal of the Geological Society* (1848), vol. IV, p. 70.

part in the formation of the valleys than would now be conceded, his views of the relationship of the drainage to the dome-shaped uplift are otherwise in accordance with those of later writers.   The reader will find references to the papers of these in my Presidential Address already cited (pp. lxx and lxxi).

The dome is not absolutely symmetrical, but rather elliptical, with its longer axis trending east and west.   The ellipse also is rather narrower at the east, so that the uplift may be compared to an old caddy-spoon with its handle at the east end.   In addition a want of symmetry in the Skiddaw area may indicate the occurrence of a subsidiary dome.

The cause of the uplift is doubtful.   I have suggested elsewhere that it may be of a laccolithic character, but it is of course possible that it is simply due to earth-movement unaccompanied by intrusion of igneous rock beneath.   The subject is further considered in Part II of my Presidential Address (p. lxxxv).

After the production of this dome, and as a consequence of the formation of the relatively elevated ground due to the movement that produced it, the erosion of the Lake District, which has largely caused its present features, took place.

The radial drainage of the district, as above described, has been claimed as a case of "superimposed drainage," having been originally imposed on the newer rocks which have now been removed by erosion. This would explain the fact that the sources of the rivers do not coincide with the axis of uplift of the rocks which now appear at the surface,— an axis which as we have seen passes through the Skiddaw Fells in an east-north-east and west-south-west direction.   The present streams are distributed with no regard for that axis, as would naturally result if they were, as argued, initiated by another uplift the centre of which lies much further south and with which the sources of the consequent streams do actually coincide.

# CHAPTER XV

## EVENTS IN THE DISTRICT BETWEEN THE TIME OF FORMATION OF THE DOME AND THE GLACIAL PERIOD

The uplift of the dome would cause the initiation of a series of radial rivers flowing seawards in the general directions of the radially-dipping strata of Carboniferous and later date, which at the time of that uplift formed the higher portion of

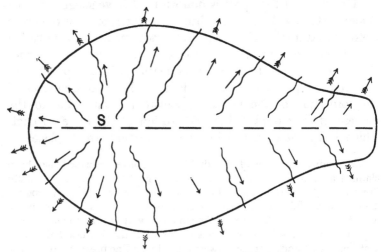

Fig. 21. Diagram illustrating radial drainage.

Outer line shews margin of "Caddy-spoon" area.
Dotted line indicates main watershed.
Wavy lines represent rivers, with unfeathered arrows shewing direction of flow.
Feathered arrows shew dips of Carboniferous rocks.
*S* Scawfell.

the dome. If the dome were symmetrical, these rivers would be truly radial, but on account of the caddy-spoon shape there would be departure from a truly radial distribution of the rivers, which would be radial over the "bowl" portion of the "spoon" and more rectilinear over the "handle" portion. This is what we actually find. (See Fig. 21).

There are no less than seven important valleys starting within two or three miles of Scawfell, the central point of the district, and radiating outwards. These are, starting at the south-east, Langdale-Windermere, Duddon, Eskdale, Wastdale, Ennerdale, Buttermere-Crummock, and Derwent. Of the valleys starting from the ridge of the "handle" of the spoon we find the Ullswater, Haweswater and Swindale vales on the north of the axis; and Troutbeck, Kentmere, Long Sleddale and Borrowdale (not the well-known valley of that name) on the south: the rivers in some of these valleys have had their courses modified as the result of subsequent change, but the present trend of the rivers approximates to their original trend and clearly brings out the dependence of the drainage system upon the dome-shaped elevation.

The rivers thus initiated would erode these valleys and tributaries would develop in the normal manner, in accordance with the well-known laws controlling river erosion. At first the valleys would be carved out entirely in the newer strata overlying the Lower Palaeozoic rocks; the valleys would be at first comparatively shallow and narrow, and the higher ground left between them extensive. At this time the topography of the district would not be very diversified.

By degrees the rivers would cut through the covering of newer rocks in places, and expose the older rocks on the valley floors. This would probably occur first in the case of the main streams, but by degrees the tributaries would also cut down into the older rocks. If we could produce a geological map of that time we should see the general surface of the district covered by the newer rocks, with the older rocks exposed on the valley floors of main streams and tributaries, giving tree-like outcrops of their junction with the upper rocks above. In time erosion would occur to so great an extent that the older rocks would also be exposed on the passes separating river from river. As this went on, the newer rock would occur in isolated patches on the uplands, and these patches would be separated by an intricate network of outcrops of the Lower Palaeozoic strata. Such a condition of things may now be seen in Bohemia to the south-east of Prague, where the river valleys occur in Lower Palaeozoic rocks, which crop out on their floors and far up their sides, while the plateau summit between the valleys is yet occupied by thin masses of Carboniferous and Cretaceous date which have hitherto escaped final removal by erosion.

As time went on these patches of rock over our district would further be reduced in size by continuance of the erosive action: one by one they would disappear until at last all were removed. In the meantime the main line of junction between the continuous mass of newer rock and the older rocks of the dome would also, as the result of erosion, be gradually receding further from the centre of the district, and this recession is of course still proceeding by the cutting back of the scarps of the newer rocks which face inward towards the district.

When the whole of the newer rocks had been removed, a general geological map of the district would present the appearance of the actual map.

When the rivers were still running over the former cover of newer rocks, tributaries would be developed in the normal manner: subsequent streams would course over the softer rocks along the lines of strike, and these would in turn receive minor tributaries.

After the newer rocks were removed the rivers would necessarily have to adapt themselves to the changed conditions with considerable complication of the pre-existing drainage system, for the streams would now tend to divert portions of their courses along the softer belts of the older rocks. There being comparatively few belts of strata of sufficient softness to allow of development of strike-streams, few of these have been formed on a large scale. The principal soft rocks are those of the Ellergill division of the Skiddaw Slates, and on a smaller scale the Skelgill beds of Valentian age. An important strike-stream along the Ellergill beds is seen east of Keswick: there the lower waters of the Glenderamackin, continued downwards as the Greta, flow along the strike of these strata.

A number of strike-streams follow the course of the Skelgill beds, as Browgill, Holbeck Gill (Skelgill), Pull Beck west of Windermere, Yewdale Beck Coniston, Mealy Gill and Appletreeworth Beck, southwest of Coniston. The last-named is the most important of these streams.

Several strike-valleys run along the junction between the Lower Palaeozoic and Carboniferous rocks, causing the gradual recession of the escarpments of the latter rocks. Among these is part of the

Lowther and its tributary, west of Shap, the Birkbeck on the south side of Shap Fells, part of the Ehen west of Ennerdale Lake, and to the south the streams which flow past Cartmel, and along the foot of the escarpment of the limestone to the west of Kendal.

In the general absence of belts of softer strata, the more important shatter-belts have played the chief part in causing changes in the drainage. A number of streams have been diverted along these belts, or initiated afresh in them. Among such are the streams flowing down Long Sleddale, Kentmere and Troutbeck[1], but others of various degrees of importance are scattered through the district and are specially marked among the rocks of the Borrowdale Series, where these belts are so well developed, and so markedly contrasted as regards durability with the rocks which they traverse.

The new conditions produced important changes in accordance with the law of unequal slopes. When streams flowed along softer rocks or along shatter-belts, erosion occurred to a greater depth than in the case of other streams which flowed over rocks of greater durability, and accordingly the tributaries of the former streams captured the headwaters of the latter. An interesting case occurs near Coniston. The Tilberthwaite Valley has a stream which once flowed to Langdale, and its waters were ultimately discharged into Windermere. The old stream course is marked by a very low col in the middle of a wide valley, close to Tilberthwaite hamlet. The waters now turn at right angles near the hamlet and are discharged into Yewdale, and so into Coniston Lake.

This was due to the low level of Yewdale Beck flowing over soft Silurian rocks close to the point of capture, whereas the original stream ran over several miles of resistant rocks of the Borrowdale Series before reaching the softer Silurian rocks. It will eventually be seen that the actual diversion was due to glacial interference: before glacial times the capture had almost but not quite been accomplished.

A case due to a similar cause is seen at Honister Pass. The stream flowing to Buttermere reaches the soft Skiddaw Slates sooner than that which flows into Borrowdale, and the upper waters of the latter stream have been captured by the former. In walking up from Borrowdale to the head of the pass, a combe is seen between Grey Knotts and Fleetwith Pike, forming the head of the valley. The tributary streams in this combe descend in the fan-like manner characteristic of combe-streams, but though most of them flow into the feeder of the Borrowdale drainage the most easterly turns sharp round through the great gash which forms the Pass, and flows towards Buttermere.

Numerous similar diversions might be mentioned, but as it is not always clear in the present state of our knowledge how far, if at all,

---

[1] See Marr, J. E., "The Waterways of English Lakeland," *Geographical Journal*, 1896, vol. VII, p. 602.

these have been produced or assisted by glacial action, it is unnecessary to describe them, and we will confine our attention to one of them, as it will be noted in a later chapter on account of an interesting feature connected with it. At Dunmail Raise, a stream flows westward from Helvellyn and turns sharply southward at the Raise to flow into Grasmere. The pass is slightly nearer to Thirlmere than to Grasmere, but as the former lake is more than 300 feet higher than the latter, the slope towards Grasmere is greater than that towards Thirlmere. Accordingly the stream flowing to Grasmere is cutting back at its head, and capturing streams which at one time flowed into Thirlmere.

The above remarks are only intended to indicate some of the changes produced by a necessity for re-adjustment, when the streams after flowing over newer rocks were imposed upon those of Palaeozoic date. A connected account of the changes remains to be written.

We have now to consider the diversified topography of the district. In doing so we must take into account several factors, of which the principal are:

(1)   Amount of uplift;
(2)   Influence of the rocks;
(3)   Nature of the eroding agents;
(4)   Time during which erosion has taken place.

The amount of uplift is important as affecting the energy of the erosive agents upon the land. If the land were flat, water would be stationary and no erosion could occur, and, other things being equal, the greater the amount of uplift the greater the velocity of the streams and therefore the greater their erosive power. Uplift then was necessary to "set the mill in motion."

Before the glacial period it may be taken as probable that the main agents of erosion were the action of the weather and of running water in the form of streams. We are therefore at present concerned chiefly with the second and fourth of the factors mentioned, though we must refer incidentally to some effects of the first.

It is a well-known law of erosion that, other things being equal, the hard rocks tend to resist erosion, while the softer ones are worn away, and that accordingly on a surface originally fairly level, the hard rocks will stand out as eminences, while the softer ones form relative lowlands.

A very cursory glance at the Lake District will shew that the three types of Lower Palaeozoic rocks, namely, the Skiddaw Slates, Borrowdale Series, and Silurian rocks, differ much as regards hardness, and give rise to very different types of scenery, so that any one standing near the junction of the Skiddaw Slates and the Borrowdale Series will at once see the contrast between the smooth, usually grass-clad hills composed of the former rocks, and the generally craggy type of scenery of those of the latter. Again, standing near the junction of the Borrowdale Series of rocks and the Silurian deposits, the tamer scenery of the latter is at once contrasted with the bolder type of the former. These differences in the general aspect of the scenery depend therefore upon the nature of the rocks of the three great groups of the Lower Palaeozoic formation.

But it soon becomes apparent that as regards relative height, the influence of hardness of the rocks is not always apparent on a large scale. We find that the softer Skiddaw Slates on Skiddaw occur at a height of 3054 feet, only 156 feet lower than Scawfell Pike (3210 feet), the highest hill composed of rocks of the Borrowdale Series. Again, though the Silurian Slates only rise to a height of 1819 feet in Lakeland, they reach 2220 feet in the Howgill Fells to the south-east. The apparent independence of height on the durability of the strata in these cases requires an explanation. The most probable (which is supported by other facts) is that erosion did not start from a nearly level surface, nor yet a simply curved one, but that subsidiary domes occurred in the Skiddaw tract, and in the Howgill Fell region beyond the margin of the district, and that the rocks were raised to an exceptional height (in the case of Skiddaw approximately to that of the main dome) by these subsidiary uplifts. Erosion would go on for some time before the cover of newer rocks was removed. When the plane of junction separating these from the older rocks was reached erosion would be checked among the harder rocks (for the Skiddaw Slates themselves are more resistant than most of the rocks of the Carboniferous and later dates). Since the period when this line of junction was reached, time sufficient only for the removal of a slight amount of the older rocks on the hill summits may have elapsed, hence such hill summits would roughly define the variations of height of the folded plane of junction. If erosion proceeds in the future as it has done in the past, we may expect that after sufficient time has elapsed, the Skiddaw Slates will be worn to lower levels than the rocks of the Borrowdale Series on the hill summits, and that ultimately the summit of Skiddaw will be at a much lower level than that of Scawfell.

The present difference in the scenic characters of the tracts occupied by the rocks of the three great divisions is in fact probably not so much due to difference of hardness between the three types of rocks, as to the difference between the uniform characters of the Skiddaw Slates, and to some extent of the Silurian rocks, and the very varied characters of the volcanic rocks, and also in a very considerable degree to the greater regularity and larger scale of many of the divisional planes in these volcanic rocks as compared with those of the Skiddaw Slates and Silurian strata.

Many of the marked contrasts between the general scenery of the three types of rock have no doubt been emphasised by the erosion in glacial times, but it will be convenient to consider those contrasts here. In addition it will be well to say a few words concerning a fourth type of scenery which is typical of the Mountain Limestone hills of the borders of Lakeland.

The Skiddaw Slates consist very largely of argillaceous rocks. There is, it is true, much grit, but the grits, with local exceptions, are fine-grained. Grit and shale alike are generally jointed with innumerable minute joints, and accordingly break into small pieces as the result of weathering action. The clay-rocks become separated into their component particles as the result of the mechanical effects of weathering, and by solution of the binding material between the grit grains, the gritty rocks are also marked by separation of their component grains. Thus slates and grits alike become reduced at the surface into a mass of fine incoherent material which masks the solid rock beneath on the flatter ground, and by being carried

down as rainwash, and by the general downward creep of the
material, tends in time to cover up such cliffs as may have
been formed previously. This loose material furnishes a soil
favourable for the growth of moorland vegetation, as grass
and heather; hence the generally smoothed outlines of the
Skiddaw Slate district at the present time; and as we shall see,
such smoothed outlines must have been even more conspicuous

Fig. 22.   Derwentwater, Bassenthwaite and Skiddaw.

shortly before the advent of the glacial period.   The illustrations
(Figs. 2, 22 and 32) shew the nature of the scenery of the
Skiddaw Slate tracts.

Among the rocks of the Borrowdale Series we find very
different conditions.  There is frequent alternation of more
and less durable rock, the lavas tending to resist weathering
to a greater extent than the ashes, when the latter are unaltered.
When they are altered into the flinty type of rock they resist
weathering in a degree equal to that of the lavas. The resistant

rocks are often of considerable thickness. Furthermore they are traversed by larger and more regular joints than are the Skiddaw Slates.

The rocks become detached in large slices along these joints, and though extensive screes develop at the bases of the cliffs thus produced, if the resistant rock is of considerable thickness it takes a long time to cover a high cliff with rock waste. Hence for longer periods than in the case of the Skiddaw Slates, we should meet with alternate cliff and slope, such as now characterise much of the ground occupied by the rocks of the Borrowdale Series, and also no doubt characterised it in all but the latest stages of pre-glacial erosion.

As the result of changes in glacial and post-glacial times, a similar alternation of cliff and slope occurs among these rocks in many parts of the district at the present day. I have already called attention to the terraced character of the hills between Thirlmere and Derwentwater, rising up to Bleaberry Fell; this is the result of alternation of lavas and hard volcanic breccias with softer and more yielding ashes, all of which over the greater part of these fells have a gentle dip.

The cliffy structure is also brought out well among the nearly horizontal beds of altered ashes and breccias of the central part of the district in the Scawfell group of Fells, and on Langdale Pikes. (See Fig. 23.)

Sometimes, as previously stated, the cliffs are less regular, of small horizontal extent, and of a general elliptical shape. In these cases, as already suggested, a nearly horizontal movement has occurred on a small scale and the crushed rock along these belts of movement has readily weathered, giving rise to ground between the cliffs covered with detritus. This mode of occurrence of the cliffs is the normal one over the greater part of the district occupied by the volcanic rocks in which cliffs occur.

The regular joints and tear-planes of the Borrowdale Series have allowed of a much greater development of shatter-belts in the rocks of this series, than in the Skiddaw Slates and the Silurian rocks. These shatter-belts are planes of weakness along which weathering and erosion by running water take place more readily than in the surrounding rocks. Gullies are formed on the hill-sides along these belts, and where the belts stand up against the sky-line, a little depression is formed along the line of the belt which forms a notch at that line. Hence the outlines of the hills formed of the rocks of the Borrowdale Series often shew a markedly crenellated character, when they have not been smoothed over by other changes. The plexus of dykes which penetrates these rocks in the centre of the district adds to this effect. Many of these dykes weather more readily than the surrounding rocks, giving rise to similar

Fig. 23. Langdale Pikes.

notches, often on a large scale: for instance Clifton Ward states that Mickledore, between Scawfell Pike and Scawfell, owes its existence to a dyke which passes through it. Characteristic examples of the scenery of the tracts occupied by the rocks of the Borrowdale Series are shewn in Figs. 24 and 40.

Occasionally the lavas exhibit a columnar structure, of which a good example is shewn on Mardale Ill Bell (Fig. 25).

The scenery of the Silurian tract presents in some respects characters intermediate between those of the Skiddaw Slates and those of the Borrowdale Series, though approximating more closely to the former. The hills are much lower than those of the Skiddaw Slates. This is undoubtedly due in some degree to the fact that the part of the district where Silurian rocks occur was not elevated, when the dome was formed, to the same extent as that portion where the Skiddaw Slates are

Fig. 24.   Scawfell.

found. But this does not altogether account for the relative lowness of the Silurian ground. There is in most places a marked and sudden diminution of height, when we pass from the rocks of the Borrowdale Series on to the Silurian strata, as well shewn along the tract between Troutbeck and Appletreeworth

Beck; it is especially clear to the south-west of Coniston village, where the Silurian rocks form a low terrace backed by the heights of Old Man, Doe Crags, Walney Scar and Caw.

The comminution of the Silurian rocks has taken place in much the same way as that of the Skiddaw Slates, but alternations of grit and slate are more marked, and the grits often give rise to prominent features. It has already been noted that the Middle Coldwell beds stand out as a ridge. The Coniston

Fig. 25. Columnar structure in lava: Mardale Ill Bell.
A detached column is seen in the foreground.

Grits, of greater thickness than these, generally form a wider ridge, while the Bannisdale Slates, with rapid alternation of grit and slate, give rise to very hummocky ground, with innumerable small cliffs which are most marked where the gritty constituent is in excess. As already noted (p. 6) the view of the country at the foot of Windermere (Fig. 4) well illustrates the nature of the area occupied by Silurian rocks.

The intrusive igneous rocks on a large scale are often resistant, and give rise to eminences, where the softer surrounding rocks have been removed by erosion.

Carrock Fell stands up as a prominent hill, and the precipitous east end especially is due to apposition of the hard gabbro and granophyre to softer rocks of the Skiddaw Slate series.

Many smaller masses of intrusive rocks amongst the Skiddaw Slates produce similar eminences, as Castle Head, near Keswick. Ward remarks that dykes on Skiddaw Dodd form well-marked buttresses which add greatly to the picturesque character of that conical hill.

The Skiddaw granite, exposed only in valley bottoms, has no appreciable effect upon the scenery, and though the micro-granite of St John's Vale and the Shap granite produce outstanding hills, the effects are not very marked.

The Ennerdale granophyre on the other hand does rise into marked eminences, which, though resembling to some extent those of the volcanic rocks, differ in detail.

The Eskdale granite is marked by rather uninteresting hummocky ground, not often rising to great elevations. In pre-glacial times it was probably weathered into tors, which, as noted by Bernard Smith, have been destroyed as the result of glacial erosion.

The effects of the smaller dykes are variable, according as they resist erosion in a higher or lower degree than the surrounding rocks. In the former case they stand out as ridges, which are not usually of sufficient height to produce a marked effect; in the latter they are worn into gullies, often deep and narrow, and these, as stated elsewhere, like the shatter-belts do give rise to features which, though on a subsidiary scale, often have a very marked character.

The Mountain Limestone country has a character of its own, marked by long scars and gentle slopes, the escarpments and dip slopes of the usually gently inclined strata. This structure is often known as desk-structure, recalling the appearance of a writing-desk on a large scale.

A prominent character of the rocks is due to their colour, a whitish-grey, which causes the hills to possess a whiteness, forming a marked contrast to the darker hues of the more ancient rocks. The best illustrations of this type of scenery are found around the estuaries to the south of the district, where the blending of the effects of the white scars, the peaty estuarine tracts, with the estuarine waters further seaward, and the Lower Palaeozoic hills of Lakeland proper rising in the background, affords scenes of singular beauty. Scout Scar and Underbarrow Scar, Whitbarrow, Arnside Knott, and Warton Crag illustrate the type.

The view of Warton Crag is an admirable illustration of the desk-structure (Fig. 26).

An interesting feature of these limestone hills is the effect of chemical solution of the limestone by acidulated water. It is well known that the limestone is soluble in water charged with carbonic acid and certain acids derived from decaying vegetation. Accordingly the general upper surface of a limestone tract becomes lowered slowly: at the same time the water acts along the divisional planes, causing the enlargement of the joints by solution. These are converted into gaping fissures or grikes, often penetrating many yards below the surface. As the dominant joints often lie at right angles to one another, one finds a fairly flat surface of limestone traversed by a rectangular network of grikes. Such surfaces are known as clints. They are well seen on Kendal Fell, Cartmel Fell, and in many other places. Where the limestone is thin-bedded the acidulated waters work along the bedding planes, forming horizontal fissures leading inwards from the grikes, and when two of these meet a surface slab of rock becomes detached and loose. Every minor fissure and depression is acted on differentially by the water, and accordingly the upper surfaces of the clints are often covered with fantastically shaped blocks of limestone corroded into every variety of knob and hollow; these blocks are familiar from their frequent use as rock-work in gardens.

In places streams have been swallowed up by these grikes, and working along bedding planes have formed caverns in the limestone. Limestone caverns of this type have been found in Warton Crag, and have furnished relics of animals.

No doubt the formation of those clints and grikes which we now see has taken place since the glacial period, but as

Fig. 26.   Warton Crag near Carnforth;  shewing "desk-structure."

similar structures must have been formed before that period, it is convenient to consider them in this place, while we are discussing the general effects of the rocks upon the scenery of the area.

We must now consider time as a factor in controlling topographic features.

During the earlier stages of erosion of a newly-formed land-tract the topography is comparatively simple, as the river valleys are at first shallow and narrow, and the intervening tracts between the valleys retain their primitive characters.

As erosion proceeds the valleys are gradually deepened and widened, by the transference of weathered material from the slopes into the rivers. The tracts between valley and valley thus become narrowed, and if the change goes on for a sufficient length of time the intervals between the valleys are reduced to ridges.

Ultimately the main streams establish their base-levels and can erode downwards no more. At this period the diversity of topography is most marked, the difference of level between ridge top and valley floor being at its maximum. The tributary streams continue to erode, when the downward erosion of the main streams has ceased: weathering continues on the valley sides and ridge summits, and the latter are gradually lowered. The slopes become lessened, bare rock is gradually covered by an accumulation of waste, and the ridges and hill summits along those ridges lose their sharpness, become rounded, and their rocks also become masked by a covering of waste-material. The topography during this period undergoes simplification, and a tract of rounded waste-covered hills separating flat-floored graded valleys results. A hill-tract of this type is spoken of as one of subdued relief.

It is essential to the right understanding of the changes which occurred in the district during the glacial period that we should recognise that before that period the hills of the district were in a state of subdued relief, and we must now consider the evidence bearing upon this.

If we study an area of subdued relief, which for some reason has not been greatly modified by glacial action, the topography of the period of subdued relief still remains with little change. Such a tract occurs in the immediate vicinity of Lakeland, constituting the Howgill Fells, as shewn by Prof. Fearnsides and myself[1].

Though many of the topographic features of the Lake District have undergone great modification during the glacial period, we still find abundant traces of this relief remaining among the various Lower Palaeozoic rocks. It is well shewn

Fig. 27.  Round-topped hill near Hartsop.

in some of the Skiddaw Fells at the back of Skiddaw, and the hill on the right-hand side of the accompanying illustration (Fig. 27) shews an example amongst the volcanic rocks of the Borrowdale Series. This is a hill above Hartsop, near the head of Ullswater. Even when parts of a hill or hill-range have undergone subsequent modification, the outline of subdued relief is frequently well marked on the sides facing south or west, or having an aspect between those two points. This may be noticed on the High Street and Helvellyn ranges, and on that of the Coniston Fells. The section in Fig. 28, taken from Kentmere to a hill west of Dunmail Raise, which is drawn to

[1] "The Howgill Fells and their Topography," *Quarterly Journal of the Geological Society* (1909), vol. LXV, p. 587.

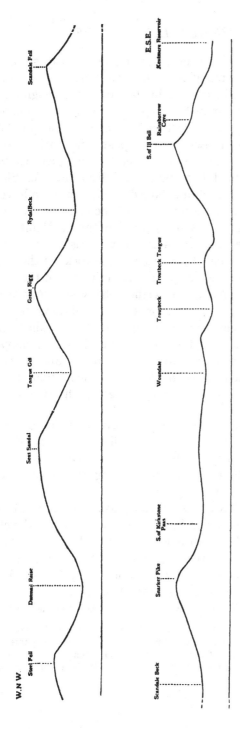

Fig. 28. Section from Kentmere Reservoir to Steel Fell, west of Dunmail Raise. (True Scale.)

The right-hand portion of the upper part of the figure is continuous with the left-hand portion of the lower part.

true scale, shews this feature. It is perfectly clear, since we get in the district every modification from the hill practically shewing the outline of subdued relief in its entirety, through others in which it has been partly destroyed in an ever-increasing degree, to that in which its destruction is complete, that all the hills of the district once possessed this topographic feature, and that its partial or entire destruction is due to subsequent changes[1].

It may be noted here that the hills formed of Skiddaw Slate, during the period of subdued relief, appear to have been more pointed than those composed of the rocks of the Borrowdale Series. This seems to be due to the contiguity of the streams on these impervious rocks, whereas they were more widely separated among the more pervious rocks of the Borrowdale Series. There was time therefore in the case of the Skiddaw Slate hills for the narrow watersheds between the streams to be reduced to ridges by transport of material down the slopes, whereas in the wider uplands separating the valleys in the rocks of Borrowdale age, the time was insufficient in many cases to allow the uplands to be converted into ridges, and they largely possessed flattened rounded slopes at their summits. This would apply equally to the hill-tops along the ridges.

Before passing on to discuss the events which occurred in glacial times, attention must be called to an event which, although it need not directly have affected Lakeland, certainly produced important modifications along the coast bordering it on the east and west.

It is known that around many parts of Britain are buried valleys, the bases of which are often far below sea-level, and that they are filled with glacial drifts. They date therefore from a time prior to the accumulation of these drifts in the valleys. One such valley has been found in the Furness district, at Park House mine, near Dalton. Here, in a narrow strip of ground, the drift was found of exceptional thickness,

---

[1] The subdued outline of the British hills in pre-glacial times was first shewn by Prof. W. Morris Davis. An admirable paper shewing this, and the principal subsequent modifications due to glacial action, will be found in the *Scottish Geographical Magazine* for 1906 ("The Sculpture of Mountains by Glaciers").

the greatest amount observed being 537 feet. As these mines are
not very far above sea-level, there is indication here of a buried
valley at least 450 feet below that level. This valley was
formed by subaerial erosion, and we here get evidence of the
last great earth-movement of the district, which was one of
depression. Its effect was to drown those valleys which were
not filled with drift above present sea-level, thus giving rise
to the marginal estuaries of Lakeland, as those of the combined
Irt, Mite and Esk, and of the Duddon, Leven and Kent rivers.

The exact date of this submergence is doubtful; it need
not have been immediately pre-glacial, indeed there is some
evidence elsewhere that the country at the beginning of glacial
times was only a few feet lower than at present. All we can
say is that before the filling-in of these valleys by drift, the
land was at a considerably higher level than its present one,
and at some subsequent date it was depressed to its present
position[1].

# CHAPTER XVI

## THE GLACIAL PERIOD: THE ICE-SHEET

Between Liassic times and the beginning of the glacial
period, many geological ages intervened, which are not
represented by any deposits in the district or its immediate
neighbourhood: accordingly we have been compelled to
attempt an account of the events which occurred during those
ages without the assistance of any deposits.

When we reach Pleistocene times we are once more able to
utilise deposits in unravelling the history of the district, for
many glacial accumulations are found in and around it.

Before describing these deposits and the other evidences of
glacial action, it will be advisable to offer a few general remarks
upon the glaciation of the district.

Shortly after the researches of Agassiz on the effects of

---

[1] Accounts of the Park Mine borings will be found in the *Geological Survey
Memoir*, "The Geology of the Southern Part of the Furness District in North
Lancashire," p. 4.

glaciers, Dean Buckland brought forward evidence to prove that glaciers had in past times occupied the uplands of Great Britain. In the second portion of his classical paper, published by the Geological Society in 1840[1], he gave instances of the work of glaciers in Lakeland, noting the moraines and the rounded, polished and striated surfaces of rock. One of these he describes as occurring in Dr Arnold's garden at Fox Howe, near Ambleside. He also attributed the remarkable distribution of the boulders of Shap Granite, which had already been observed, to the action of ice. This paper by the Dean is the starting-point of those many researches on the glaciation of Britain, which are still far from complete.

For some time after Buckland's paper was written the glacialists were divided in opinion as to the relative importance of land-ice and floating sea-ice as agents of glaciation, and for some areas in Britain the matter cannot be yet regarded as ultimately settled.

Furthermore, at the time of Buckland's researches, the effects of glaciers of the Alpine type alone had been studied in detail. Subsequently the effects of ice action on a much larger scale have been examined, and we know now that some areas, of which Greenland is the typical example, are at the present day glaciated, not merely by valley-glaciers of Alpine type, but by a great ice-sheet, which in the interior of the country covers hill and dale alike beneath an icy pall.

The formation of a similar ice-sheet in the upland regions of the north of England was ably advocated by R. H. Tiddeman[2] in a paper which may well be regarded as marking the beginning of a new epoch in the study of the glaciation of the north of England. Tiddeman's views were supported in a paper by J. G. Goodchild dealing with the district immediately north of that studied by Tiddeman[3]. Clifton Ward, in papers which

---

[1] Buckland, W., "On the evidences of Glaciers in Scotland and the North of England," *Proceedings of the Geological Society*, vol. III, pp. 332, 345.

[2] Tiddeman, R. H., "On the Evidence for the Ice-Sheet in North Lancashire and adjacent Parts of Yorkshire and Westmoreland," *Quarterly Journal of the Geological Society* (1872), vol. XXVIII, p. 471.

[3] Goodchild, J. G., "The Glacial Phenomena of the Eden Valley and the Western Part of the Yorkshire Dale District," *ibid.* (1875), vol. XXXI, p. 55.

will be subsequently referred to, gave the results of his detailed studies of the glaciation of the Lake District. He regarded floating ice as an important factor in producing some of the effects of glaciation in the district. This would require a submergence of the district to the extent of some 2000 feet below its present level, and, as will be argued, there is no sign of such submergence. The evidence brought forward by Tiddeman and Goodchild for the existence of an ice-sheet to the east and south-east of the district is equally apparent in the district itself, and it is now generally held that the district during the period of maximum glaciation lay under an ice-sheet. While rejecting Ward's views of submergence, one must pay a tribute to the accuracy of his observations on the glacial phenomena. Lake District geologists owe a great debt of gratitude to Ward for his work, not only on glaciation, but on a number of other subjects connected with the geology of the district.

Let us consider briefly what must happen during a period of glaciation, of which the maximum stage is sufficiently long and severe to give rise to an ice-sheet.

The beginning of glaciation would be marked by the development of small local glaciers in the upland valleys of the hill districts. As glaciation increased these glaciers would creep down towards the lower ends of these valleys, and at the same time the ice would increase in thickness. As the ice extended lower and lower down the valleys it would go on rising up the valley-sides, until two glaciers would at places become confluent over depressions in the intervening ridge. As this rise went on more and more of the ridges would be buried beneath the ice, until ultimately valley, ridge and hill-top alike would be buried. At the same time the ice would ever be extending its outer limits, and if the glaciation were sufficiently severe might ultimately extend beyond the borders of the uplands on to the surrounding low ground.

The greater erosive effect of an ice-sheet of this type would certainly destroy at any rate most of the effects of glaciation produced by the valley-glaciers of the earlier phase, and accordingly in a district like Lakeland we can hardly expect to find any traces of this early and comparatively feeble glaciation which nevertheless must certainly have occurred.

After the maximum stage had been reached, changes would occur in inverse order to those which marked the waxing of the ice-sheet. The ice would begin to shrink backwards at

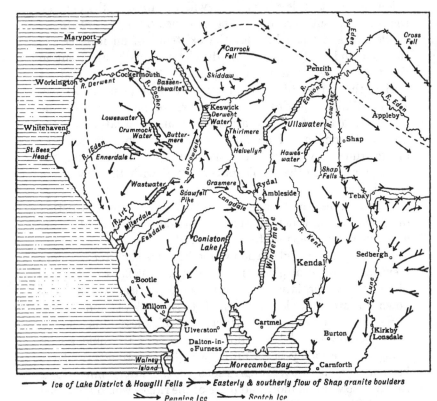

→ *Ice of Lake District & Howgill Fells* ⇒→ *Easterly & southerly flow of Shap granite boulders*
↠ *Pennine Ice* ⟹→ *Scotch Ice*
⤻—×— Limit of north-easterly and easterly distribution of Shap boulders.
- ˵ ˵ Southerly limit of Scotch Ice.

Fig. 29. Map shewing directions of movement of the Ice-Sheet.

its margin, and diminish in thickness. By degrees the hill-tops and ridges would emerge through the waning ice, and the ice-sheet would be replaced by valley-glaciers, which would ultimately shrink until the final stages would be marked by the occurrence of little glaciers in the combes beneath the mountain summits. The effects of these final stages of glaciation, unlike those of the initial stage, would be preserved, and might modify to some extent those produced during the period of maximum glaciation.

It may be regarded as certain that the Lake District passed through these phases during the glacial period, but it is by no means certain that the phases were of the simple character here outlined. It is indeed most probable that the general waxing and waning of the ice were complicated by periodical advance and retreat on a small scale, due to minor and shorter climatic variations. What is not clear is whether such changes assumed so great an importance that periods of waning produced a practical or entire disappearance of the ice between two or more periods of intense glaciation: whether in fact the glacial period can be separated into sub-periods of an alternate glacial and genial character.

As we can hardly expect to find evidence of the work of the local glaciers before the development of an ice-sheet, it will be well to discuss at the outset the evidence which was originally relied upon as proving the existence of such a sheet. This evidence is largely connected with the direction of the glacial striae, and the tracing of boulders of ice-transported rock to their parent sources. The accompanying map (Fig. 29), founded on such evidence, is largely based upon the observations of Clifton Ward, Goodchild, and D. Mackintosh, though the facts observed by others have also been utilised[1].

In studying this map it must be taken into account that the movements of the ice in the upland valleys, as indicated by direction of striae and transport of boulders, are to a large extent those of the valley-glaciers during the period of waning

[1] Ward, J. C., "The Origin of Some of the Lake-Basins of Cumberland," *Quarterly Journal of the Geological Society* (1874), vol. xxx, p. 174, and "The Glaciation of the Southern Portion of the Lake District...," *ibid.*, 1875, vol. xxxi, p. 152. See also "The Geology of the Northern Part of the English Lake District," chap. xiii.

Goodchild, J. G., *loc. cit.*

Mackintosh, D. Several papers in the *Geological Magazine*, 1870: also "On the Drifts of the West and South Borders of the Lake District...," *ibid.*, 1871 pp. 250 and 303, also papers in the *Quart. Journ. Geol. Soc.*, 1873 and 1874, vols. xxix and xxx

of the glaciation. It is probable that the lower parts of the ice-sheet during the period of maximum glaciation moved in the same general direction, being restricted by the valley-sides, but the upper part of the ice would in many cases move transversely to this: evidences of such movements may be occasionally obtained on the ridge summits in the heart of the district, but the evidence of the movement of the ice-sheet is more clearly exhibited when we reach the lower ground upon the outskirts of the district.

It must also be noted that though for the sake of simplicity a definite line is drawn between the ice-sheet of the Lake District and that which occupied the south of Scotland and came into contact with the Lake District ice, that line is arbitrary, and indicates only in a general way the most southerly extension of the Scotch ice at its maximum. The intermingling of boulders of Lake District and Scotch rocks on the low ground to the north of Lakeland, and also to the north-east over parts of Edenside, shews that there was debatable ground, occupied by Lake District and Scotch ice alike. This may be due to alternate occupation, or to the over-riding of one ice-sheet by another, or to a combination of both causes. It is perfectly clear that at times the two sheets were confluent, as indicated by the resultant movement of the Lake District ice towards the south-west and south on the western side of the district, and to the south-east on the eastern side. It is clear that the Lake District ice could not move westward on the western side, though, unless a barrier had existed, it should have done so over the low ground of what is now the Irish Sea, and should also have moved in a northerly and north-easterly direction over the low ground of Edenside. Instead, it is deflected along the line of the Cumbrian coast on the west, and up the Eden Valley and over the Pass of Stainmore into Yorkshire on the east.

As regards the general movement of the ice which had its origin within the confines of the district, it will be noticed that the main ice-shedding line coincides with that of the present main watershed, being in the direction of the axis of the caddy-spoon-shaped dome, and having a general east and west trend.

It extends from the Scawfell Hills in the west, past the Rossett Gill and Stake Passes, and over Dunmail Raise, the Grisedale Pass, Kirkstone, High Street, and so to Shap Fells on the east.

It will also be noted that an ice-sheet on the Howgill Fells has affected the movement of the Lake District ice, causing one lobe to move eastward up Ravenstonedale and over into the Eden Valley, and another southward from the Shap Granite over some minor ridges and valleys into the Kent, and down the low-lying ground towards the mouth of that river to Morecambe Bay. To the south only, the ice of the district seems to have been unchecked by any barrier, and to have made unimpeded progress far southward over Morecambe Bay and the low-lying land of Lancashire to the east of that Bay.

Such is the general nature of the distribution of the ice-sheet and of its directions of movement, as indicated by the distribution of the glacial striae and the transport of the various boulders. We may now proceed to consider this in greater detail, devoting special attention to the directions of travel of the boulders.

In all the larger and often also in the smaller valleys of the greater portion of the district, the direction of the striae and the travel of the boulders shew that the lower parts of the ice, as already stated, moved down the valleys. But on the ridges we occasionally find evidence of the oblique passage of the ice across the directions of the main valleys, indicating that the upper parts of the ice on many occasions moved in a direction differing from that of the lower parts confined between the valley-sides.

Some cases in illustration of this statement may be given. They are taken from Clifton Ward's "Geology of the Northern Part of the English Lake District." It is in this part of Lakeland alone that a detailed examination of the direction of the scratches and transport of the boulders has been made over a considerable tract of ground. No doubt similar cases will be observed in other parts of the district, and indeed this has occasionally been done.

Ward's observations were made when mapping the country which is represented on the Ordnance Survey one-inch map of the Keswick district. We may consider the different ridges from east to west. In the east we have that between Thirlmere and Borrowdale, in the centre that between the latter valley and the Buttermere-Crummock Vale, and west of this the ridge between that vale and Ennerdale. It may also be noted that in the district shewn towards the north-east of the map, there is evidence that the ice coming from the Helvellyn range went across the depression in which is the railway from Keswick to Penrith. A lobe of this ice went some way northward up the higher part of the Glenderamackin Valley, as shewn by the occurrence of boulders of the Armboth and Helvellyn dyke in the neighbourhood of the hamlet of Mosedale. On the ridge between Thirlmere and Borrowdale, ice-scratches are seen, as north of Blea Tarn, with a general north-north-west direction, that is, oblique to the ridge and pointing towards Borrowdale. Again boulders of the Armboth dyke were found near the top of Bleaberry Fell at a height of about 1800 feet. These must have travelled in a direction somewhat west of north, even if derived from the most westerly outcrop of that rock. Ward concludes therefore that "the ice in the Thirlmere valley was of such thickness and so pressed against the western side by the great supplies off the long Helvellyn range, that...it partly escaped across the western watershed south of Armboth Fell."

Between Borrowdale and Buttermere, boulders of volcanic rock are found resting on Skiddaw Slates at the summit of Maiden Moor on the top of the ridge between Derwentwater and the Newlands Valley at a height of 1887 feet. This hill is half a mile north of the nearest outcrop of the volcanic rocks. From thence they are traceable all along the ridge to the top of Cat Bells, so that the ice here reached the ridge-summit. From the fact that at one place a train of boulders can be traced uphill from the low ground at the head of Derwentwater to Hause Gate at the summit of the ridge, it would appear that some of the ice from Borrowdale must have escaped over the ridge into the Newlands Valley. To the north of this there is evidence of movement of the ice from the Derwent to the Cocker Valley between Whinlatter and Wythop.

Turning now to the ridge between Buttermere and Ennerdale, we find a very remarkable occurrence recorded by Sedgwick and Ward, who referred it to the action of floating ice. Starling Dodd on the watershed is about a mile west of Red Pike. This Pike is composed of granophyre, while the summit of the Dodd is of Skiddaw Slate. On the summit of the Dodd however are boulders of the granophyre, which rock is only found at a higher level than that of the Dodd on Red Pike itself. We here get evidence of the transport of boulders in a westerly direction right across the ridge at a height of over 2000 feet.

It is clear from the facts noted that the ice rose to heights of over 2000 feet, and that the ice masses of adjoining valleys in some cases became confluent over the intervening ridges. Considering the proved thickness of the ice, and the heights of many of the ridges, which are often below the 2000 foot contour-line, we must expect to find such evidence frequently, when a more detailed examination of the glacial phenomena has been made. Indeed, there is even now plenty of evidence, though in most cases, as we should expect, the movement of the ice was generally parallel with the trend of the ridges: it is only when the ice of one valley was more powerful than that of the adjoining valley that the movements across the ridges which have been briefly noticed could occur. It may be noted that Bernard Smith has proved that the ice went over the top of Black Combe (1969 feet).

Cases of deflection of ice on a larger scale have been ascertained, especially where masses of ice originating beyond the confines of the district have come into contact with Lake District ice. To appreciate these, it will be necessary to consider the general lines of travel of the boulders of the more important and easily recognisable rocks, and this we may now do.

The most striking feature in the distribution of the boulders is the frequent occurrence of rocks of the Borrowdale Series on ground composed of Skiddaw Slates to the north, and of Silurian rocks to the south, and the practical absence of any boulders of these Skiddaw Slates and Silurian rocks upon the ground occupied by the Borrowdale Series. This is sufficient to prove a general outward movement of the ice from the central part of the district, a movement the nature of which is fully confirmed when we study the distribution of the boulders in detail. To the north of the district the volcanic boulders are found scattered over the north Cumbrian plain. To the north-east and east they are traceable over the Eden Valley and over Stainmore far away into eastern England. To the south they are scattered profusely over the southern flatter parts of the district; thence they were carried over Morecambe Bay, and along the west Lancashire plain into the heart of the Midlands.

Apart from the rocks of the volcanic series, those which are most useful for detecting the lines of movement of the ice are the various intrusive igneous rocks, supplemented by observation of the boulders of metamorphic rocks which occur in their aureoles. As the right understanding of the movement of the ice which originated in Lakeland itself is dependent upon our knowledge of the movements of ice which approached the confines of the district from other regions, we may first consider the distribution of boulders of Scottish origin around the margins of the more northerly parts of the district. The most useful rock for this purpose is the granite of Criffel which is readily distinguishable from the granites of the Lake District. Boulders of this granite are found on the west, north and north-east sides of the Lake District. Their distribution on the west has been described by Mackintosh. They are traceable along the Cumbrian coastal region from the Solway to the neighbourhood of Ravenglass, and have subsequently been followed southward to the neighbourhood of Millom.

On the north and north-east sides of the district their distribution is indicated by Goodchild. The boulders are thickly scattered over the north Cumbrian plain, having been transported over the Solway. They are then traced between the line of their south-western limit, shewn on the map, Fig. 29, and the western slopes of the Pennines, up the Eden Valley, past Appleby and Brough and finally over Stainmore. This Scotch ice acted as a barrier, preventing the free escape of the Lakeland ice to the west, north and north-east.

We may now proceed to consider the dispersal of boulders derived from Lake District rocks.

The boulders from Carrock Fell have been carried in an easterly direction, and then, affected by the movement of the Scotch ice, they have travelled in a south-easterly direction up the Eden Valley past Appleby. Some of the rocks of the Skiddaw granite probably pursued the same direction, though I am not aware that they have been traced. Clifton Ward traced the boulders from the granite exposure in Sinen Gill, near the head of the Glenderaterra Valley. They moved southward down the valley, and at its junction with that of the Greta went in a westerly direction to Keswick, and coming within the influence of the ice moving down the Derwent Valley, were then carried northward.

Boulders of the Threlkeld micro-granite carried by the ice from the Helvellyn range were stopped from a direct northerly movement by the opposing mass of Skiddaw and Saddleback. They are distributed in a crescentic manner around the south side of that mass. Thickly spread over the ground of Matterdale Common, one horn of the crescent extends towards Bassenthwaite to the north-west, and another towards Trout-beck to the north-east. In the latter tract some have gone a little way up the Glenderamackin Valley with the boulders of the Armboth-Helvellyn dyke, and the others nearly due east. As the latter are found on Great and Little Mell Fell they must have passed over the watershed east of Troutbeck Station into the area of the Eden Valley drainage.

The distribution of the boulders of the Sale Fell lamprophyre is remarkable; as Ward has shewn, these boulders extend for some miles in a direction a little west of south, that is from the lowlands towards the uplands. Ward attributed this to the action of floating ice, but the movement is in accordance with what has been observed in the Eden Valley, as the result of the general southerly movement of the Scotch ice over the lowlands on the northern side of the district: at one time a lobe of the Scotch ice must have extended southward as far as the upper end of Bassenthwaite Lake and the country to the west of it.

The distribution of the boulders of Buttermere and Ennerdale grano-phyre is also of great interest. As this rock occurs on the north and south of the main watershed of the district we find that boulders have travelled northward and southward. On the north, one train of boulders extends from the northern end of the outcrop of the rock through the Vale of Cocker. As the ice which carried them reached the low ground north of Cockermouth, it came within the influence of the Scotch ice, the margin of which must then have been north .of its position when the Sale Fell boulders were transported. Along this tract we meet with an intermixtuie of Lakeland and Scotch boulders. The united ice moved eastward, on the north side of the uplands of the district, and along this tract we find boulders of the granophyre carried into the Eden Valley, and some way to the south up that valley. The stream of boulders carried down Ennerdale again came within the influence of the Scotch ice on reaching the low ground, and moved southwards, as did also that which came down Wastdale. Accordingly we find boulders of this granophyre along the southern part of the Cumbrian coast, at first occurring on the whole to the east of the Scotch boulders, but further south, as the Scotch and Lake District ice mingled, the boulders from the two countries were stranded over the same belt of ground. From thence the boulders of the granophyre were carried far south over the lowlands east and south of Morecambe Bay and into the Midlands.

The boulders of Eskdale granite, like those of the Ennerdale granophyre, were at first carried outwards from the hills towards the sea, down the valleys of the Irt and Esk, but they also were directed southwards when the lowlands were reached, and travelled in that direction along with the Ennerdale boulders far beyond the limits of Lakeland. It is interesting to notice that on reaching the south end of Cumberland a lobe of the ice extended eastward over the low ground by Millom, and was even forced for some distance up the Duddon Valley, as proved by the occurrence of boulders of Eskdale granite at Dunnerholme, on the east side of the Duddon estuary.

The last rock, from which boulders have been dispersed, requiring more detailed consideration, is the Shap Granite. The remarkable character of this distribution has been the subject of many memoirs, the earlier of which date from a period anterior to the discovery that glaciation had affected this country. At that time the distribution of the boulders was regarded as having been caused by violent floods. Detailed study of the distribution of the boulders may be dated from the appearance of a paper by Prof. John Phillips[1]. The tracts of ground over which these boulders have been scattered in the vicinity of the district will be seen on the map (Fig. 30). There are two main areas where the boulders are found, one being generally east and the other south of the granite outcrop. The eastern development starts from the granite, and passing over the ridge of Carboniferous rocks between

[1] Phillips, J., "On the removal of large blocks or boulders from the Cumbrian Mountains in various directions" (1837), *Report of the British Association for* 1836, p. 87.

Ravenstonedale and the Eden Valley, extends in a fan-like form over the Edenside lowlands as far north as Melmerby. Its eastern limit is defined by the barrier of the Pennines, against which the ice was banked. The boulders are traceable as far south as the latitude of the Stainmore depression. Further south the Lake District ice was checked by that coming from the hills at the head of the Eden Valley, and was forced over Stainmore carrying the granite boulders into Teesdale, and ultimately to the east of Yorkshire. On the south side, the movement of the Lake District ice was checked to the east by the ice of the Howgill Fells, and the boulders are only traceable down the Lune Valley for

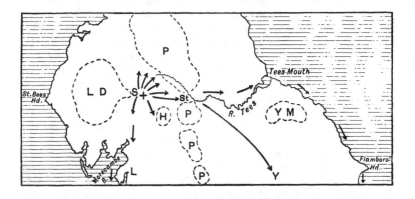

Fig. 30.　Map shewing distribution of Shap Granite Boulders.

| | | | |
|---|---|---|---|
| S | Shap. | LD | Lake District. |
| L | Lancaster. | P | Pennines. |
| Y | York. | H | Howgill Fells. |
| St | Stainmore. | YM | Northern Yorks Moors. |

two or three miles south of Tebay. Further west however a lobe of ice from Shap went over cols into Borrowdale and Crookdale and ultimately the boulders found their way into the valley of the Kent, down which they were carried to Morecambe Bay, and thence southwards towards and into the Midland Counties[1]. This southerly stream of boulders can hardly have moved southwards at the same time that granite boulders were being carried eastward and north-eastward. They no doubt mark a period when the influence of the Scotch ice was at its maximum, causing the deflection of the ice issuing from the eastern valleys of the Lake District in a southward direction. This is borne

[1] See a description of this train of boulders in the *Geological Survey Memoir* by Dr A. Strahan, "The Geology of the Country around Kendal, Sedbergh, Bowness and Tebay," p. 47.

out by the striated rocks at the foot of Haweswater, on which some of
the striae point down the valley, while others on the same rocks are
at right angles to the former, and run parallel with the outer margin of
the high ground of the district in a direction a little east of south.   As
the former are fainter than the latter, it looks as though the valley ice
coming from the head of the Haweswater Valley was overcome at a later
period by ice from Scotland and north-eastern Lakeland.   Goodchild
also notices the striae on Shap Fells sweeping in a curve with a general
north and south trend, its concave side facing the high ground.   The
intermingling of erratics in the lowlands of Edenside on the strip of
ground where boulders of Scotch and Lake District rocks occur together
suggests the occupation of that belt by ice from the two areas at different
times, just as does the intermingling of erratics which we have noted
on the north and north-west sides of the district.

# CHAPTER XVII

### THE GLACIAL PERIOD: ICE-EROSION

Having considered the development of the ice-sheet over
the Lake District, it remains for us to discuss its effects on the
erosion of rock, and the accumulation of the eroded material.

Erosion no doubt occurred during the preliminary and
final stages of glaciation, but its effects must have been most
marked during the period of maximum glaciation, and we may
therefore consider the effects of erosion at this stage.

That a certain amount of ice-erosion has taken place has
long been known, and indeed was proved by Buckland in his
classical paper.   Every upland valley, where the rocks are
sufficiently durable to preserve the effects, shews evidence of
such erosion; the rocks are rounded, with the rounded portions
facing up the valleys, while those which face down-valley are
often rough and irregular.   These rounded *roches moutonnées*
are furthermore frequently polished and striated, and everyone
now admits that these marks are signs of glacial erosion.

I give in Fig. 31 an illustration of one of these *roches moutonnées*, occurring in the Bannisdale Slates near Bowness-on-Windermere, shewing the rounded part facing up-dale and the rough part down-dale.

For a long time it was maintained that these changes were all that were produced by ice-action in the way of erosion. It was believed that the ice had simply removed the asperities of projecting rocks, and that no great modification of the topography of a district had been caused by ice-erosion. Of

Fig. 31. *Roche moutonnée* near Bowness-on-Windermere.

recent years the effects of such erosion have been shewn to be much greater than this. A very extensive literature on this subject is available, much of it bearing upon the glaciation of the British Isles. The major effects of glaciation were independently described by two writers, and I shall apply their conclusions in considering the evidence of erosion in the Lake District. The papers are by Dr Harker and Prof. W. Morris Davis[1]. Harker's paper deals with the island of Skye; Davis' is a general application of the results which he had previously

[1] See Harker, A., *Transactions of the Royal Society of Edinburgh* (1901), vol. XL, Part II, and Davis, W. M., *Scottish Geographical Magazine*, February, 1906.

attained, to the glaciation of a district in which the hills had been already reduced to outlines of subdued relief, as was the case with those of the British Isles.

These authors have clearly shewn that the slight modifications produced in the formation of *roches moutonnées* are quite subsidiary to the extensive widening and deepening of the valleys of the upland regions, and that this enlargement of the valleys has produced profound modifications in the topography of districts like our own, where glaciation has occurred in a high degree.

It will be convenient if we consider at the outset the various modifications of pre-existing features which are now regarded as indicative of glacial erosion, and afterwards point out the dominant instances of such features in the Lake District. Before considering these modifications, however, it is desirable to give a fuller account than has already been given of the topographic features of hill and dale in an area of subdued relief which has not been subjected to glacial erosion.

The nature of the mountain summits and ridges has already been considered. The mountain-tops and ridge-summits are characterised by more or less rounded outlines, covered with waste material. This waste extends down the valley-sides, covering the rocky cliffs of an earlier stage of water-erosion, and reaches to the valley-floors. The valleys are widely opened and have well-graded floors from source to termination, devoid of waterfalls and lakes. The tributary streams are graded to the main streams and do not enter in a series of falls or cascades. The rivers in most cases swing from side to side of the valley, so that on looking up the valley a continuous view of the valley-floor is not obtainable, as a series of over-lapping spurs proceeding from alternate sides of the valley hides the valley-bottom.

These features, which Davis gives as characteristic of a hill district in the stage of subdued relief, which has not been subsequently glaciated, may be found slightly modified when glaciation for any reason has not been pronounced. I have already alluded to the characters of the Howgill Fells, which have been only slightly glaciated, and where the pre-glacial features therefore exist with only slight modifications. Parts of the Skiddaw tract of fells exhibit similar features, probably owing to the same cause, namely, that the ice originating in those fells was prevented from obtaining a free outlet, owing to the presence of ice from the Helvellyn range, from the Derwent Valley, and from Scotland. In the central part of the Lake District, where the ice was powerful and capable of free outward movement, the modifications are very great, and it is in the tract of country occupied by the rocks of the Borrowdale Volcanic Series that most of the effects of glacial erosion are exhibited in the highest degree.

Before considering these effects further, let us revert for a moment to consideration of the *roches moutonnées*. As before stated, it was argued that the rough ends of these prominences facing down valley were due to their having been unaffected by glacial erosion, and that they actually represented pre-glacial surfaces. We find plenty of evidence in the district, as elsewhere, that the rough faces, as well as the smoothed portions, have been produced by glacial erosion, but in a different manner, the rough faces being due to plucking away of rock-masses by the ice utilising divisional planes, while the smooth surfaces are due to grinding action. If the reader will examine the figure of the *roche moutonnée* (Fig. 31) he will notice some joint-planes crossing the rock and will see that plucking has taken place along these joints. This feature is admirably shewn in many of the upland valleys: it is specially well exhibited in a series of *roches moutonnées* in Borrowdale in the neighbourhood of Quay Foot Quarry.

We may now consider the differences between the effects of ice and running water as affecting the topography of a district, and it may be remarked that, apart from effects due to the physical characteristics of ice and water respectively, these differences are to some extent of degree only and not of kind.

In the first place a glacier occupying a valley is wider and deeper than a river, and if it erodes, its bed will also be wider and deeper than the river-bed. Accordingly one would expect a simplification of form in the ice-modified valley. Lateral erosion would tend to destroy the ends of the projecting spurs, truncating them so as to produce triangular facets at right angles to the direction of the movement of the ice. Vertical erosion would deepen the glacial bed, and the result of these effects would be the production of a straightened and smoothened valley with steep sides, and a canal-like cross-section. The ice at the valley-heads would similarly modify those heads into half-bowl-shaped forms, with steep sides, forming combes.

Mr Willard D. Johnson[1] has discussed the influence of the *Bergschrund* in accelerating the erosion in the case of combe-glaciers, and the effects of these great crevasses must be reckoned with as a factor in producing the modifications which give rise to combes.

The effect of steepening the valley sides and heads would be to produce a state of unstable equilibrium, causing excessive downward transport of weathered material, often in the form of landslips. This downward transport would continue after the disappearance of the ice, and is now going on, tending to smooth over once more the steep slopes produced by glacial erosion. It would be especially effective during the waning of the ice, when the hill-tops and higher ridges became exposed, and were subjected to intense frost action.

In connexion with the deepening and widening of the valleys and the formation of combes, an important feature should be noticed, namely the greater effect of glaciation on hill-slopes facing between north and east than on those between south and west. This is marked by the greater abundance of combes facing the former directions, and also by the asymmetric outlines of ridges, to which attention has already been directed. There are many local exceptions, but they are too few to modify the general rule that in the glaciated regions of the world the combes on the whole are on the shaded sides, and that the asymmetric ridges are also more profoundly glaciated on the shaded sides.

G. K. Gilbert[2] has suggested that the production of these asymmetric forms is due to the production of longitudinal crevasses along the shaded side, owing to greater thickness of ice on that side, and a consequent tendency to flow across the glacier. At the bottoms of these crevasses he believes that erosion would take place similar to that advocated by Johnson in the case of the *Bergschrund*.

[1] Johnson, W. D., *Journal of Geology* (Chicago, 1904), vol. XII, p. 569.
[2] Gilbert, G. K., *Journal of Geology* (Chicago, 1904), vol. XII, p. 579.

I would venture to suggest another possibility. On the sunny side, melting of the ice might take place to so great an extent that the proportion of morainic material held in the ice became large enough to check the corrasive power of the ice, which would therefore corrade laterally and downward on the shaded side only.

As the ice in the main valleys would be greater in volume than in the tributary valleys, and as furthermore the ice in the latter would tend to be ponded back by that of a main valley, the erosion in the main valley would be greater than in the tributaries, and accordingly the floors of the tributary valleys would no longer enter the main valleys at grade, but after the disappearance of the ice, the brooks from the tributaries would cascade down the smoothened sides of the main valleys to reach the main river. Tributary valleys of this nature are known as *hanging* valleys.

Davis points out that the action of ice here is like that of rivers, but on a large scale. The *surface* of the tributary glacier would be graded with that of the main glacier, just as the surface of a tributary stream entering at grade into a main river flows tranquilly into it without marked change of slope. But even in the case of water-streams in this condition the bottom of the bed of the tributary is at a somewhat higher level than that of the main stream, and if the stream were dried up, one would see the bed of the tributary forming a miniature hanging valley.

Hanging valleys are produced otherwise than by glacial erosion, but there is one marked feature in those produced by glacial erosion, namely, that in several cases the side-streams of water flow down into the main valley without having cut a bed for themselves. Now if the U-shaped or canal-shaped outline of the main valley had been produced by water-erosion, it would require a very considerable time to produce this outline, and during that time the tributary streams, where entering a main valley, should have carved a gorge in the side of the main valley. In some cases they have done so, but usually in a slight degree; in others however there is practically no defined bed to the cascading tributary, as seen, for instance, in the stream coming down from Giller-combe to the Seathwaite Valley near the head of Borrowdale, and that which descends from Bleaberry Tarn to Buttermere.

These hanging valleys will not only be produced when a tributary enters the main valley, but where minor tributaries enter more important ones, so that if we walk up a main valley and follow the tributaries in the order of their importance to the culminating watershed, we should pass the mouths of several of these hanging valleys in succession.

Let us now consider the longitudinal profiles of the glacially-modified valleys.

The bed of a graded river has a steadily diminishing slope down stream.  This, as shewn by Harker, is not the case with an ice-eroded valley, when erosion will be greatest where the thickness of the ice is greatest, as, for instance, where the valley has re-entrant or concave forms.  Accordingly, within limits, the steady gradient, which for water-erosion is the stable form, is for ice-erosion *un*stable, since any departure from it leads to a further departure.  Therefore the longitudinal profile of a valley affected by ice-erosion "consists of two or three stretches of moderate slope divided by relatively steep drops, over which the water cascades."  In the stretches of moderate gradient we may even find "a negative or reversed gradient," if the ice has excavated more deeply up valley than further down; there will, of course, be an upward slope for some distance down valley, giving rise to a rock-basin, which when the ice recedes will contain a rock-bound lake.  Lakes of this character are to be regarded as only minor inequalities bearing a small proportion to the general lowering of the valley-floor by glacial erosion.

Here also, as pointed out by E. C. Andrews, we find a similar process on a small scale in the case of water-erosion, where the general downward grade of the bottom of the thalweg is modified by the erosion of small pools when water-action is at its greatest.  In other words water, like ice, can erode to some extent when moving up-hill, though only on a small scale.  The action is well exemplified in the rock-pools which occur below waterfalls and cascades.

While recognising the efficacy of moving ice as an erosive agent, attention may be called to an aid, the full significance of which has probably not yet been sufficiently appreciated.  The plucking action of ice seems to demand the existence of inequalities to allow of the initiation of this plucking effect.  There is a very considerable amount of subglacial water beneath the ice, as shewn by the emergence of glacier streams from the snouts of glaciers. Much of this water no doubt flows along the bottoms of the valleys, but this stream will receive tributaries, and these tributaries may in places flow over depressions in the spurs which occur in a valley before its modification by ice-erosion. They would cut troughs across these spurs, and the effect of these troughs according to Professor O. T. Jones would be to produce intervening upstanding portions which would be prone to removal by plucking action of the ice. Prof. Jones has kindly allowed me to refer to this inference before he has published it.

An effect of subglacial waters plunging down crevasses to join the streams at the glacier-bottoms is the wearing of potholes on a large scale in the rocks of the glacier-floor. These giant-kettles resemble in many respects the potholes formed by ordinary river action around cascades, but they are on a larger scale, and often occur in places where the action of ordinary rivers is out of the question. They are, no doubt, usually concealed by drift, but a few have been detected in the district, as on the ridge between Rydal and Grasmere; and I think it probable that the "Kail Pot" near Howtown on Ullswater was produced in this way, though I have not recently examined it.

We may now refer to some of the more important illustrative examples of the various features described, as developed in the district.

It is unnecessary to enumerate the canal-shaped valleys: all the main valleys and the greater number of the tributaries shew the canal-like cross-section in a marked degree. It is indeed rare to find a valley which existed in pre-glacial times having overlapping spurs that have only undergone slight modifications as the result of glacial erosion, though some occur in parts of the district composed of Skiddaw Slates.

In the centre of the district ice-erosion has taken place (accompanied by rock-weathering above) to so great an extent that the original rounded mountain tops and ridges have been converted into peaks and sharp arêtes. Some of the hills of the Scawfell group, including Scawfell Pike and especially Bowfell, have been converted into peaks. Of arêtes we may notice those on Saddleback, and on the east face of Helvellyn.

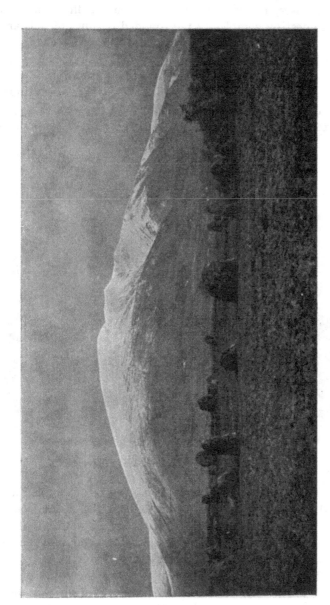

Fig. 32.   Truncated Spurs, Saddleback.

In many cases, modification has not gone on in a sufficient degree to destroy completely the rounded outlines of hills and ridges. In these cases there is usually a marked change of slope between the gently inclined, unmodified ridges and the steep, lower valley-sides affected by glacial erosion.

Truncated spurs are abundant. They are beautifully shewn on the south side of Saddleback. In the view (Fig. 32) these steep slopes are indicated by the absence of snow upon

Fig. 33. Burtness Combe, Buttermere.

them. Falcon Crag, Borrowdale, is a good example, and Bernard Smith notices their occurrence in the Black Combe district. They occur with such frequency, however, that an enumeration would be tedious. It will be good practice for the visitor to the district to detect them for himself.

Combes are extremely abundant among the rocks of the Skiddaw Slates and Borrowdale Series, though the most striking are found among the latter rocks. The illustration shews one known as Burtness Combe, on the west side of

Buttermere (Fig. 33). Those of the Coniston Fells, and the fells above Haweswater, are among the best examples. Some of these combes are short hanging valleys, lying laterally upon the sides of larger valleys, as for example Gillercombe in Borrowdale and the Bleaberry Tarn combe near Buttermere. The former is really a hanging valley with a combe at its upper end. Others are found at the heads of main valleys, and the steep slopes beneath are apparently of the nature of steps produced by unequal glacial erosion. On either side

Fig. 34. Combes, Langdale Pikes.

of the Stake Pass, between Langdale and the Greenup Valley, occur combes which form the heads of those valleys, and are separated from their lower stretches by steep descents. The combe above Warnscale Bottom at the head of the Buttermere Valley is probably of this character, as also that west of Nabs Moor at the head of Swindale Beck, a tributary of the Lowther. In the two latter cases the main valley goes behind the combe and the waters descend in cascades to the level of the lower tract of the valley, so that the cliffs below the combes may be of a composite character, being of the nature of steps steepened by lateral erosion.

In some cases we find one combe below another. This is seen on the north-east side of Coniston Old Man, and there is a good example on the south side of Langdale Pikes, as shewn in Fig. 34. Here the combe of Pavey Ark is at a high level, with a smaller combe still higher on its western side. The stream from Stickle Tarn at the foot of Pavey Ark descends down the side of a lower combe whose floor is at no great height above that of the main valley.

Fig. 35. Asymmetric Ridges.
(Head of Windermere below.)

These double combes are also probably due to the formation of steps along the longitudinal course of the glacier, but the steepness of the steps, as for instance below the Low Water combe on Coniston Old Man, suggests complications.

It is possible that in some stages in the formation of these combes, arranged in line at various levels, the lower combes may have been occupied by *remanié* glaciers, which while eroding caused the recession of the cliff above to its present steep slope.

Examples of asymmetrical ridges may now be noted. The High Street and Helvellyn ranges have a general north and south trend. In each of these cases the slopes on the western side are largely those of subdued relief, while those on the east side are concave curves due to more profound glacial modification; in addition the present combes face eastward.

The asymmetry of ridges where unmodified by combe formation is particularly well exhibited in the ridges coming from the Helvellyn group towards Windermere and Grasmere on the south.   Anyone viewing these ridges from the western side of the head of Windermere will see this type of topography to perfection.   Four ridges are seen between the head of Troutbeck and Dunmail Raise, namely, those of Red Screes, Scandale Fell, Great Rigg, and Seat Sandal.   The two former are shewn in Fig. 35.

Fig. 36.  Asymmetric Ridge.
Looking down Kentmere.

The diagram on p. 141 (Fig. 28) shews the features.

It will be observed that on Seat Sandal and Great Rigg, the concave curve produced by glacial modification does not reach the fell-summits, which retain the rounded outlines of subdued relief.   In the case of Scandale Fell, it has cut past the summit, which is therefore pointed.   Though Snarker Pike, along the line of section, shews the rounded outline, Red Screes, standing further back, has the same features as those of Scandale Fell.   The illustration (Fig. 36) is a view of the ridge between Kentmere and Troutbeck, which is seen at the extreme right of the diagram (Fig. 28).   In figure 36, which is a view looking southward, the concave slope is towards the left and the convex one to the right.

In the Scawfell group glacial modification has largely destroyed the pre-glacial features in the higher regions, though they are noticeable on some of the lower slopes.

The Coniston Fells, stretching north and south, again shew curves with convexity on the western sides, and those with concavity on the eastern. Here again the topography is complicated by the abundance of combes. These are some of the most striking examples of a feature which is almost universal over the district.

There are certain exceptions to the rule that the concave slopes face between north and east. A prominent case is Saddleback, where the subdued outlines face northward and the combe topography is on the south (Fig. 32), and again on Illgill Head above the Wastwater Screes the concave curve (here unmodified by combes) faces westward and the convex outline of subdued relief eastward. In these cases we have evidence of exceptionally powerful ice-currents on the sides which possess the concave curves.

Hanging valleys whose streams enter the slopes of the main valleys some way above their floors are very numerous. The best examples are found in those parts of the district occupied by rocks of the Borrowdale Series, though they also occur in the Skiddaw Slate tract to the north, and, as stated by Bernard Smith, among the rocks of that age on the Black Combe Fells. Many of them are quite short, being indeed mere combes, but others have a greater length. The accompanying illustration (Fig. 37) giving sections through the combe containing Bleaberry Tarn, and through Gillercombe, shews the nature of these valleys.

Towards the east of the district, Measand Beck runs through a well-graded valley east of Haweswater and descends to that lake in a series of cascades. The combes at the head of the Haweswater Valley are probably due to the formation of longitudinal steps. On Helvellyn we find hanging valleys on the eastern and western sides. The combes containing Red and Keppelcove Tarns are at the heads of valleys, and appear to be of the nature of longitudinal steps, but four lateral

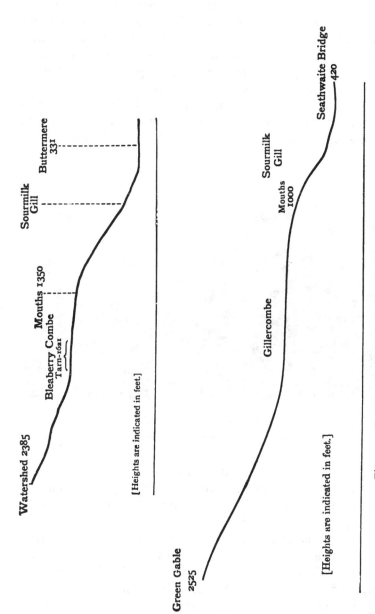

Fig. 37.  Sections through Bleaberry Combe and Gillercombe.

combes are seen on the western side of Grisedale.   Another set
of combes on the west of Helvellyn opens laterally on to the
sides of the Thirlmere Valley.   These combes extend from near
Dunmail Raise to Great Dodd.   The sharp summit of the east
side of the Red Tarn combe (Striding Edge) is seen in Fig. 38.
The actual combe is shewn in Fig. 45 (p. 186).

Some of the most striking valleys of the type we are considering are found in Borrowdale.   Lowdore Falls are due to
the waters of the hanging valley of Watendlath reaching the

Fig. 38.   Striding Edge, Helvellyn.
Red Tarn at the base of the combe is outside the picture to the left.

glacially steepened slope of Borrowdale.   Towards the head
of Borrowdale we find the Greenup Gill combe hanging as
regards the Langstrath Valley.   But the most interesting
hanging valleys are in the Seathwaite branch, which starts at
Esk Hause, and is well graded from near this point downwards.
As one descends, one sees on the left the waters of the hanging
valley in which Sty Head Tarn is situated falling over the
steep cliff as Taylor Gill Force.   North of this is the very
beautiful example of the Gillercombe stream descending the
slope as Sourmilk Gill.   (This is a term which we find

elsewhere, as for instance below Bleaberry Tarn Buttermere, and Easedale Tarn Grasmere; where we meet with it we may know that a hanging valley occurs.)

Still further north the stream of the graded valley from Honister Pass descends by the cascades of Hause Gill. A little lower down the main valley and on its opposite side the stream from Combe Gill descends a steep slope to enter the main valley near Stonethwaite.

Fig. 39.  Hanging Valleys in Borrowdale.
As seen from path above Rosthwaite.

Taking the hanging valleys which enter directly into the main valley in order, and starting from the head, we have Taylor Gill, Gillercombe, Honister, Combe Gill and Watendlath, and it is interesting to notice that the streams from these begin to descend the slope into the main valley at lower levels as we pass down that valley. The view (Fig. 39) shews Combe Gill on the left with the stream cascading from its mouth, Taylor Gill in the middle, and Gillercombe on the right.  The stream from the latter is concealed behind the nearer hill.

In my Presidential Address to the Geological Society I gave reasons for supposing that there had been some modification in the course of the Borrowdale Valley along this stretch.  If that be the case it does not materially affect the main question, though there may have been some diversion prior to the glacial period.

A series of hanging combes is found on the south-west side of the Buttermere Valley, of which the most conspicuous are those of Bleaberry Tarn and Burtness Combe already noticed (Fig. 33, p. 164).

In the Wastdale Valley some fairly well defined hanging valleys are found on its north-western side.

In Eskdale there are several of these valleys.  Whillan Beck on the north side descends to Boot in cascades.  On the south side of the valley two well-known waterfalls, Birker Force and Dalegarth Force (Stanley Gill), mark the descents of Low Birker Pool and Birker Beck from their graded upper courses.

The hanging valleys near the head of Eskdale and that of Lincombe Beck from Devoke Water are probably due to diversions of drainage, as argued elsewhere, but the cliffs have no doubt been steepened as the result of ice-erosion. That near the head of Eskdale is shewn in Fig. 40.

In the Coniston Fells, there are hanging valleys produced by steepening of the sides of the main valley.  The best example is on the west side, where we find a real valley and not a mere combe.  This is the valley containing Seathwaite Tarn, the stream from which cascades down into the flat valley of Seathwaite Beck.

On the east side of the Fells, the high level combes have already been considered, but the Red Dell is a lateral valley, as is perhaps also the valley containing Leverswater. At a lower level Church Beck, into which the various streams from the north-eastern combes and higher branch-valleys pour, itself forms a hanging valley, which flows down a fall and cascades to Coniston village.

In Langdale, besides the upland combes, Little Langdale, a valley of some length, hangs with reference to Great Langdale,

Fig. 40.  Hanging Valley.
Near head of Eskdale.

and the stream descending the slope into the latter valley forms Colwith Force.

There are a certain number of hanging valleys descending into larger valleys in the south-eastern part of the district but they are of no great interest. Woundale, which descends into Troutbeck in cascades, is possibly due to diversion.

Lastly, we must consider the longitudinal modifications with production of steps and rock-basins.

The steps are not so prominent in Lakeland as in some other British districts, but several examples occur on a small scale.

Some cases of high cliffs below combes, where the combes are situated at the heads and not at the sides of valleys, have already been noticed, and some of these are almost certainly examples of longitudinal steps. One of the best examples is seen below Grey Friars at the head of the Seathwaite Tarn Valley in the Coniston Fells. In the more lowland valleys the steps are usually of no great height. A good example is seen in the Watendlath Valley below Watendlath Tarn; another in Great Langdale below Elterwater. In each of these cases the stream cuts through the steps in falls, that in Great Langdale being Skelwith Force (Fig. 41).

The lakes are of various sizes and at very different heights above sea-level. The upland lakes in the combes are usually small tarns, while the larger lakes are confined to the low-level valleys, though in these tarns also occur.

Many lakes and tarns have their waters held up to some extent by glacial accumulations. Some are entirely so held up, others no doubt partially so, and in the latter cases it is difficult to know when we are dealing with rock-basins. There are tarns and lakes which can be definitely proved to lie in rock-basins, namely, Watendlath Tarn, Elterwater and Thirlmere, and no doubt many others will be proved to exist in the district.

Most of the larger lakes lie among the softer rocks of Skiddaw Slates or those of Silurian age, as Derwentwater and Bassenthwaite, Buttermere and Crummock, the lower part of Ennerdale, Coniston, Windermere and most of Ullswater, but

Haweswater, Thirlmere and Wastwater lie in the rocks of the Borrowdale Series[1].

Proof will be given that the surface-waters of some of the larger lakes are held up by glacial accumulations when we con..der these accumulations, but we have evidence that even in these cases the bulk of the waters at the lake-bottoms lie in rock-basins.

The sub-aqueous topography of the lake basins is a downward continuation of the sub-aerial topography: this of course would be the case

Fig. 41. Skelwith Force.

whether the basins were produced by erosion, or by subsequent "ponding" of an eroded valley, but there are special features which support the view that the larger lakes lie in basins which owe their existence *as basins* to erosion, apart from any higher rise of waters due to glacial accumulations.

In the first place some of the lakes shew separate basins below water, that is, there are deeper tracts separated by ridges. This is noticeable in Coniston, Windermere and Ullswater. The feature might be

[1] An account of the larger lakes, with maps shewing the soundings, was given by Dr H. R. Mill in the *Geographical Journal* (1895), vol. vi, pp. 46 and 135 and issued separately by George Philip and Son in the same year under the title of *The English Lakes*.

brought about by unequal accumulation of deposits in different parts, but the nature of the isobaths is more suggestive of separate rock-basins.

Again, some of the lakes are deepest near the head.   The greatest depth of Windermere (219 feet) is three miles from the head, as is also that of Ullswater (205 feet).  This is what might be expected in the case of lakes produced by erosion; if due to ponding however, it would necessitate fairly even tilting for a considerable distance if ponding were due to earth-movement, or steady downward increase in the thickness of accumulation as far as the foot of the lakes, if due to glacial deposit.

Fig. 42.  Looking down Ullswater.

One of the most cogent arguments against the formation of this type of lake by ponding was advanced many years ago by Dr A. R. Wallace.  He pointed out that lakes produced by ponding should have numerous bays where the lower parts of tributary valleys were drowned, whereas the lakes claimed as being formed by glacial erosion are characterised by the singular straightness of their sides.   This is the feature of the larger lakes of Lakeland.   Bays are few, and where they occur, there is evidence that the upper waters are due to ponding. Thus Pullwyke Bay forms a marked indentation in Windermere, which is one of those lakes shewing evidence of an origin due to blocking by glacial accumulation as well as by erosion.   The sides of Buttermere-Crummock (which once formed one lake) and of Wastwater are marked by their straightness, except where modified by delta-growth, and this is shewn in a less marked degree by the other large lakes. The feature is illustrated in the view of Ullswater (Fig. 42).

We may, therefore, claim all the larger lakes as lying in rock-basins produced by variations in the amount of erosion longitudinally down the valleys. The origin of these lakes by glacial erosion was advocated by Clifton Ward[1].

# CHAPTER XVIII

## THE GLACIAL PERIOD: ACCUMULATION

In the uplands of the Lake District the period of maximum glaciation was largely one of erosion: accumulation of material in the same area was a feature of the period of waning glaciation. The consideration of erosion and accumulation in that order is therefore in accordance with our study of events according to their sequence in time. Nevertheless, a certain amount of accumulation occurred in the interior of the district during the period of maximum glaciation, and the accumulations of this period will be considered in the first place. Speaking in a geological sense these accumulations may be regarded as having been formed contemporaneously with the erosion-features which were considered in the preceding chapter.

Ice, like water, depends for its transporting power on its rate of movement and its volume, and the rate will certainly be diminished on entering lower ground. Accordingly we find that the main accumulations of the period of maximum glaciation are spread over the lowlands bordering Lakeland; they occur in the Eden Valley, on the north-western plain of Cumberland, and around the sea-coast to the west and south. They enter into the district, being most abundant where the

---

[1] Ward, J. C., "The Origin of some of the Lake Basins of Cumberland," *Quarterly Journal of the Geological Society* (1874), vol. XXX, p. 96, and "The Glaciation of the Southern Part of the Lake District and the Glacial Origin of the Lake Basins of Cumberland and Westmoreland," *ibid.* (1875), vol. XXXI, p. 152.

ground is relatively low, as over the Silurian rocks of the southern area. In the centre and northern parts of Lakeland there are not as a general rule very extensive tracts occupied by these accumulations, though in exceptional circumstances we may find them spread over a considerable extent of ground, as for example between the northern end of the Helvellyn range and the fells of Skiddaw and Saddleback. The materials forming these accumulations consist of stony clays accompanied by sands and gravels. The stony clay is boulder clay or till, varying in character according to the material which has been ground up by ice in order to form it, but usually packed with sub-angular, often striated and polished, boulders of various sizes.

The sands and gravels are usually roughly-bedded deposits laid down by water, and they frequently overlie the boulder-clay, though patches are often found interbedded with the clay. Many of the boulders whose direction of travel we have already considered may be extracted from this clay.

Where the accumulations of this date are strongly developed, they tend to give rise to somewhat monotonous features, by smoothing over pre-existing inequalities, and as the poor soil which they form is apt to nourish a marshy vegetation, the aspect of a tract occupied by boulder-clay is usually one of dreariness, as in the case of the moorland of Matterdale Common between the Helvellyn range and Saddleback. As however these deposits do not enter largely into the uplands, the scenery of the district as a whole is not greatly affected by them.

An account of the tills and accompanying gravels will be found in Clifton Ward's memoir on the northern part of the district. A number of sections were exposed during the formation of the Cockermouth, Keswick and Penrith railway, which, though varying in detail, shew a definite succession, the lower accumulations consisting largely of stiff boulder-clay and the upper of stony or gravelly clay. These deposits have been traced up to a height of at least 1600 feet.

The till, according to Ward, "occurs every here and there in small patches among the mountains, in rock-sheltered spots, and may fre-

quently be seen in the valleys." I believe that it may have a more
extensive development than has been suspected on the higher ridges.
These ridges are often covered by vegetation and shew no sections.
They are obviously occupied to a large extent by drift of some character
and at these heights the drift is likely to belong to the period of maximum
glaciation.

I find a record in my note book of a cutting in drift which occupies
the catenary curve between Fleetwith Pike and Grey Knotts, above
Honister Pass. Here on the watershed at a height of about 1700 feet
"the material seen in the cutting is till rather than ordinary boulder-
clay."

The till often presents a comparatively level surface, at other times
it is modified into the whale-back ridges known as drumlins, the longer
axes of which have a general parallelism with the direction of movement
of the ice. They are most frequently found in the wider parts of the
valleys toward the margin of the district. Several occur in the neighbourhood of Keswick, and the islands of Derwentwater are the summits
of drumlins. Some are found around Windermere and between Windermere and the Kent Valley, and they are abundant in the lower part of
that valley.

The drift gravels which overlie the till appear to be due to the
deposition by water of materials similar to those which formed the till.
Ward refers them to marine action during submergence, but, as already
stated, we get no direct evidence of any submergence, and they are
explicable on the supposition that they were formed by outward streams
at the snout of the ice-sheet during the period of its recession.

In addition to these drift gravels there are occasional accumulations
of false-bedded sands and gravels forming mounds, known as eskers.

These eskers are most numerous in the tracts surrounding the
district, as for instance south of the Solway, and in the lowlying region
near Carnforth, but they occur in the district also, often at considerable
elevation. Ward notices several which are found in the Keswick district.
Some near Troutbeck Station occur at a height of 1045 feet above sea-
level. Others on lower ground are found in the Naddle and St John's
Vales, and in the Vale of Lorton. In the south of the district they are
also found here and there; Strahan records one at Blea Beck Old Bridge,
near Shap Wells, not far below the watershed.

These eskers were obviously deposited by water, and Ward assigns
them also to the supposed period of submergence. The fluviatile
character of eskers is now generally accepted, though the exact conditions in which eskers were formed are still a subject for discussion;
indeed there seems little doubt that they have originated in more than
one way, and until a fuller study has been made of the eskers of the
Lake District, and especially those of the marginal lowlands where

they are so much better developed, their exact mode of origin must be left in doubt.

One of the arguments advanced against submergence is the absence of marine shells in the gravels of the greater part of the district, but it is only one argument amongst many. As the question is one which demands a review of evidence obtained from a much wider extent of territory than we are now considering, we cannot enter into this question at length. Though no shells have been found in the drift accumulations of the actual Lake District, they have been found in the deposits of the adjacent lowlands, and to this occurrence we may devote our attention. At Gutterby Lane End, Gutterby Spa, near the sea-coast to the west of Black Combe, Bernard Smith records the occurrence of 80 feet of drift consisting of loamy clays, sands, gravels and boulder beds. The boulder beds contain fragments of marine shells including *Turritella communis*, *Buccinum undatum*, *Anomia ephippium*, *Ostrea* and *Mytilus edulis*. This drift, it will be observed, is in the tract of country where the boulders give proof of the encroachment of ice from over the Solway, and the shells, as in similar cases elsewhere, have no doubt been carried by the ice from the old sea-floor over which it moved.

The accumulations formed during the period of maximum glaciation are those which would increase the amount of water in the larger lakes, by adding to that which is held in the rock-basins those upper layers of water which would be ponded back if such drift accumulations were formed at the lower ends of the lakes. The diagram (Fig. 43) illustrates the formation of a lake by a combination of the processes of erosion and accumulation. At least two of the larger lakes suggest that they are due to this combination, namely, Bassenthwaite and Windermere. From the former lake, about a mile from its foot, a wide nearly dry valley extends westwards from Bassenthwaite Lake Station to Cockermouth. It is thickly covered with drift at the bottom, and a low col near Embleton Station, lying in the bottom of the valley, allows sluggish water to flow eastward and westward. At the actual foot of the lake the Derwent river issues and turns westward through a valley separated from that just noticed by a ridge rising to nearly 800 feet above the sea at Elva Hill. It would appear that in pre-glacial times the river Derwent ran through this dry valley, and that when the Derwentwater-Bassenthwaite rock-basin was eroded the valley became so choked with drift that the

waters had to find an escape elsewhere and took their present
course.

Windermere is a yet more striking case. Close to the
foot of the lake is a great bank of drift; on surmounting
this one notices a wide dry valley extending southward in
a direction continuous with that of the lake, past Cartmel
to the sea. The floor of this valley is thickly covered with
drift. The actual foot of the lake has a hook-like termination
pointing westward and at the end of this hook the waters flow
through a gap in a ridge to an adjoining valley, past Newby
Bridge and Backbarrow, to reach the estuary near Greenodd.
The river in some places is obviously too large for its bed. It
would seem to have been diverted over a low col between the
Cartmel Valley and its present course, thus accounting for the

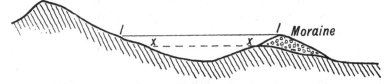

Fig. 43. Diagram of lake partly held up by moraine.
*II*    Level of actual surface of water.
*XX* Upper limit of water held in rock-basin.

hook which marks the site of a small former tributary descending
eastwards from the col, and the diversion was geologically
speaking so recent an event that the river has not yet had
time to adapt to its own requirements the floor of the minor
stream which once occupied the valley through which the
present river runs.

Let us now turn to the accumulations which were formed in
the district during the period of waning of the ice. These
naturally do not extend so far away from the centre of the
district as do the relics of the period of maximum glaciation,
and accordingly we get abundant relics in the centre of the district. Also, with rare exceptions, the higher ridges which were
covered by ice during the period of the ice-sheet were now laid
bare, and accordingly the accumulations of the later stages of

glaciation are generally confined to the valley sides and floors. These accumulations are moraines, comparable in every respect with the moraines which occur in association with the existing glaciers of an alpine region.

The moraines are mostly of the nature of terminal moraines, produced during periods of pause in the recession of the ice. Lateral and medial moraines are rarely shewn, as might be expected, for the former are often covered by subsequent accumulation of hill-waste, while the medial moraines are frequently destroyed as the result of later river-erosion. A good example of the relic of a lateral moraine is seen on the right bank of the Langstrath Valley, at the foot of Greenup Gill. The depression between it and the hillside behind has not been filled in with hill-waste. Others occur in the Scawfell district ; there is one between Angle Tarn Gill and Allencrag Gill, and a particularly fine one between Grainsgill and Ruddy Gill. The latter may be partly medial moraine.

Bernard Smith records a very fine one below Walney Scar, on the east side of the Duddon Valley. He also mentions lateral moraines on the western slopes of the southern end of the district facing the sea.

A medial moraine descends from the northern angle of Steel Fell ; it was formed between the ice which descended northward from Dunmail Raise and that which came down the head of the Wythburn Valley.

The terminal moraines are usually of a crescentic form, the convexity of the crescent pointing down the valley. They some-times form fairly continuous even-topped ridges, while at other times each crescent is formed of a series of confusedly arranged mounds. In either case, we frequently meet with a series of parallel crescents, separated by ground occupied with a thin covering of drift, if any ; this points to periods of pause during retreat of the ice.

In many cases the mounds are confusedly arranged without any apparent approach to the crescentic form.

Sections of these moraines are often exposed by subsequent stream-erosion. They consist of unstratified sand and gravel (with more clayey material among the Skiddaw Slates), containing boulders of various sizes, usually of angular form, indicating transport on the surface of the ice. The appearance of the moraines is shewn in the illustration of one at the foot of Rossett Gill (Fig. 44).

These moraines occur at various heights. The highest are found in the combes below the higher summits, and were obviously formed by corrie-glaciers. Nearly every combe in the district has its moraine, and it seems unnecessary to give examples. The confused arrangement of the moraine-mounds

Fig. 44. Moraine-mounds.
Foot of Rossett Gill, Langdale.

is well exemplified in the combe at the head of Greenup Gill, while a very perfect crescentic mound with even summit occurs below the combe of Wolf Crag behind Matterdale Common : the old bridle path from the Vale of St John to Matterdale passes along its summit. A huge regular moraine is seen in the Coniston area some way below Goatswater.

The moraines are traceable to lower levels, and frequently occur at the bottoms of the main valleys. Some of the most interesting of these are found in Borrowdale. One good crescentic moraine stretches from the corner of Stonethwaite

Fell to the hillside above Seatoller. Another very perfect one, lower down the valley, forms a continuous ridge from the corner of the same fell to the rocky knoll at Rosthwaite, against the southern side of which its extremity is plastered. The former moraine belongs to the glacier of the Seathwaite, the latter to that of the Stonethwaite Valley.

Still further down we find moraine material plastered against the rock at the end of the Rosthwaite alluvial flat: it is doubtless the relic of a moraine which once stretched across the valley, most of which has been since destroyed by erosion. This is the lowest moraine which can be traced definitely in Borrowdale, and in the other large valleys moraines are not found in their lower tracts, which suggests that the glaciation that produced these moraines had a definite downward limit.

In other cases, besides those noted near Rosthwaite, the moraines are plastered against the upper sides of rocks projecting out of the valley-floor. A very beautiful instance is seen in Swindale, the valley whose stream flows into Lowther Beck. It occurs just below the hanging valley already noticed.

A remarkable set of crescentic terminal moraines is found at the end of Bleawater Tarn, at the head of Mardale. The most important of these is traceable up the side of the steep ridge to the north of the tarn, and actually reaches its summit at Caspel Gate, a notch between Long Stile and Rough Crag, so that the corrie-glacier must have risen until it overtopped the ridge at this point and sent a lobe or ice-avalanches northward into Riggindale.

The scenery of the tracts occupied by moraines is interesting rather than beautiful: indeed these mounds, covered with the characteristic vegetation of rushes and bent-grass, with hollows between occupied by sphagnum and other vegetable growth, give an aspect of dreariness to the surroundings.

We have considered the effect of glacial accumulations upon the larger lakes; let us now regard the effects of the morainic deposits upon the tarns.

A large proportion of these tarns occur in combes, but others are found upon upland plateaux. Many of them are no doubt to some extent rock-basins, but several have their surface-waters at any rate held up by glacial accumulations. Two very beautiful illustrations of combe tarns, with moraines below, are Bleawater Tarn, the moraine of which has been noticed, and the adjacent Smallwater Tarn, but indeed the greater number of the combe-tarns shew moraines at the ends.

Blind Tarn below Brown Pike, Coniston, deserves special mention. It has a beautiful crescentic moraine around it and there is no exit from the tarn. The water soaks through the drift and oozes out upon the surface many scores of feet below the tarn-level. It is a pretty sheet of water on a combe-floor, and deserves a visit : it can be reached in seven minutes from the Walney Scar path.

Scales Tarn below Saddleback has a moraine at its end formed of exceptionally stiff clay, with many boulders, some of which shew glacial striae.

In the case of Smallwater Tarn, the old pre-glacial valley is seen below the drift accumulation, with its bottom at a lower level than that of the present stream below its exit over rock. Here we have a good illustration of what so frequently occurs, namely the diversion of the stream over solid rock, when the lowest part of the drift ridge was at a higher level than that of a little col over adjoining solid rock. The diversion on a small scale is similar to what appears to have happened on a large scale in the case of Windermere. Many cases of diversion of the outlet over solid rock are found in various parts of the district ; an example which is worth visiting by those in the neighbourhood of Esk Hause is the little High House Tarn just north of Sprinkling Tarn. The old valley is blocked by drift from Great End to the south and the waters of the tarn now have an exit down a rocky gully on the western side.

The large proportion of cases in which this diversion occurs in the streams issuing from these tarns is due to survival. In most cases where diversion over solid rock did not take place, the stream flowing over the drift basin rapidly eroded it, and drained the waters of the tarns. Countless peat-mosses behind moraines mark the sites of former tarns, which have been drained in this way. A good example is found behind the moraine below Wolf Crag, where the site of a former tarn is marked by a peat moss ;

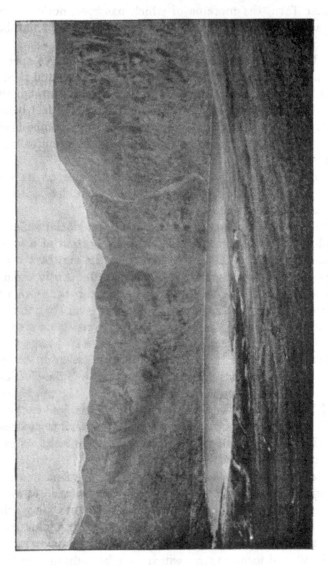

Fig. 45.   Red Tarn, Helvellyn.

the moraine at one place is gashed to its base by the stream which cut through it and drained the tarn[1].

The former lake or tarn may be rapidly converted into an alluvial flat even when diversion has caused the outlet to flow over solid rock, if the lake be shallow, and the rock sufficiently easily

Fig. 46.   Bleawater Tarn, near Haweswater.

eroded to allow of fairly rapid cutting-down at the exit from the lake.   A good example occurs in the Langstrath Valley, and was described by me some years since.   An old meander of the

[1] See Marr, J. E., "The Tarns of Lakeland," and "Additional Notes on the Tarns of Lakeland," *Quarterly Journal of the Geological Society* (1895 and 1896), vol. LI, p. 35 and vol. LII, p. 12.   One of the tarns there described, that of Watendlath, is certainly a rock-basin, and some of the others are probably partial rock-basins.

river was filled in by drift on the west side of the valley, producing a lake, now converted into an alluvial flat.   The stream from the lake flowed over rocks to the east of the drift barrier and has cut out the remarkable gorge of Blackmoss Dub, which was figured by me in a paper in the *Geographical Journal* for June, 1896.   It is also figured by Lord Avebury in his *Scenery of England.*

Illustrations are given of two tarns, the upper waters of which at any rate are held up by moraine material.   These tarns (Red Tarn, Helvellyn and Bleawater Tarn, Haweswater) are also good examples of combe-tarns (Figs. 45 and 46).

In addition to the moraine material, we find numerous isolated boulders, some of which have no doubt been derived from moraines, while others were stranded as isolated blocks. Many of these are perched blocks, standing on *roches moutonnées* and other minor eminences in such a manner that their quiet stranding is obvious : they have not fallen into their present positions from above.   Lakeland, as compared with parts of North Wales, for example, is not marked by a great profusion of these blocks, but they are sufficiently common, especially on the ridges of the higher uplands.   Many of them may be noticed in the neighbourhood of the tarns on the north side of the Scawfell group of hills.

## CHAPTER XIX

### THE GLACIAL PERIOD: GLACIAL OVERFLOW VALLEYS. OSCILLATIONS OF THE ICE

The paper by Prof. P. F. Kendall, "A System of Glacier Lakes in the Cleveland Hills[1]," marked the opening of a series of researches upon the effects of streams of water running from glaciers and ice-sheets, and from lakes held up by ice, which have subsequently been pursued in many districts.   In Lakeland, no definite examination of these phenomena has been carried on over the whole area : this is a line of study which

[1] Kendall, P. F., *Quarterly Journal of the Geological Society* (1902), vol. LVIII, p. 471.

awaits pursuit. In the extreme south-west of the district a detailed study of the phenomena has been made by Bernard Smith[1], but it is clear that similar examples are widespread in the other parts of the district. We may give a general account of the observations of Smith in the south-western district, and add a few words concerning instances of a similar kind which have been noted elsewhere.

On the western slopes of the district, between the mouth of the Esk and the Whicham, south of Black Combe, the preglacial like the present drainage was in a westerly to south-westerly direction. A series of channels is found with a general north-south trend; these are now either dry or occupied by insignificant streamlets, which obviously had nothing to do with the formation of the channels. The channels occur in parallel series at different levels, two adjacent channels being sometimes joined by cross-cuts. To understand the formation of these channels, the nature of the glaciation of this tract must be recalled. The trend of Scottish boulders from the north along the coast-line and some way inland indicates the forcing back of the Lake District ice by the Scottish ice, at the period of maximum glaciation when the two were confluent. As the ice receded, a strip of bare ground would be left between the local ice retreating to the east and north-east, and the Scottish ice retreating westward. It was the waters due to the latter which caused these overflow-channels on the coastal belt. Some are due to streams coming directly from the ice and taking the direction of least resistance. In certain cases the streams flowed between the ice and the adjoining hillside; then only half of a valley is preserved; at other times the streams left the ice margin and cut valleys which are bounded on either side by walls formed sometimes of drift, sometimes of solid rock. When the ice margin rested on a projecting spur, a notch was cut across that spur.

On the eastern side of Black Combe, and on the hilly ground to the north of it, similar channels have been cut during the recession of the ice which occupied the Duddon Valley.

[1] Smith, B., *Quarterly Journal of the Geological Society.* (1912), vol. LXVIII, p. 402.

In some cases the streams did not issue directly from the ice, but the ice ponded back the waters of the valleys, giving rise to lakes. When the lowest part of the ice barrier was higher than some col connecting the blocked valley with an adjacent one, the waters rose to the level of this col, and channels were cut from the latter to the adjoining valley. A lake of this character occupied the Whicham Valley, south-east of Black Combe. Its history is complex, but it owed its origin to the blocking of the mouth of the valley by the ice coming off the Irish Sea, and one marked overflow channel was formed at the eastern end of the lake, the stream flowing southwards towards Millom.

Those who are interested in the working out of similar phenomena should read Smith's paper, for an abstract of his conclusions, without a series of illustrative maps and diagrams, would not do justice to his labours. One point may be noticed: in the area occupied by the "Whicham Lake" he finds evidence of three beach-terraces between the hundred foot contour and a height of 180 feet. Considering the rarity of these beaches, their discovery in Lakeland is of interest, and they should be sought elsewhere in the district. Smith traces the glacial overflows of the Duddon Valley as far north as Ulpha Park, and states that he has little doubt that they would be found in similar situations as far north as Seathwaite. In my Presidential address to the Geological Society[1] I described the remarkable spur of Wallabarrow, projecting across the Duddon Valley at Seathwaite; this is gashed by two deep gorges through which the Duddon and Seathwaite Beck flow. I believe that these gashes were initiated by glacier-streams, at a place where the volume of ice of the Duddon Valley was largely swelled by the glaciers coming from the upland valleys on the western side of Coniston Old Man. There are other signs of overflow channels in the Duddon Valley above Seathwaite. Anyone wishing to make a detailed study of these phenomena in the district might well begin his labours in this valley, where the work is rendered easier than in other parts of the district, as the six-inch maps of this tract of Lakeland, which lies in Lancashire, have contours at closer intervals than have the maps of Cumberland and Westmorland.

I shall refer to one other case of overflow channels. It was previously seen (see p. 127) that in pre-glacial times, in accordance with the law of unequal slopes, a tributary of Yewdale had nearly succeeded in capturing the headwaters of Tilberthwaite Beck, which then drained into Little

---

[1] *loc. cit.*, p. cx.

Langdale. The capture was ultimately accomplished by glacial over-flow. A spur coming into the valley from the hills projects between the flat ground near Tilberthwaite hamlet and the rocky gorge down which this stream flows into Yewdale. This spur is cut by three notches.

Fig. 47. Tilberthwaite, looking eastward.

The two highest (through the upper one of which the road is carried) are dry, while the lower one, which is here a well-defined gorge separating the wider expansion of valley above and below it, is occupied by the stream. These notches have been formed during the advance and recession of the ice which occupied Tilberthwaite. This ice must have overtopped the low col which previously separated Tilberthwaite from Yewdale, and a stream flowed over it from the glacier. As the ice changed its level the stream changed its position, and finally a channel was cut to a sufficient depth to produce permanent diversion.

In the view (Fig. 47), the position of glacial overflows is seen on the right of the picture. The stream runs between the wooded part and the rock in front of it. Previous to this diversion the stream flowed down the valley in the middle of the picture, but at present there is a low col in the centre of this valley to the right of the cottages. The water beyond that col flows to Langdale.

### Oscillations of the ice.

It is clear from various kinds of evidence that the ice age was not marked by a steady waxing of ice to a phase of maximum glaciation followed by a steady waning until the ice disappeared. That fluctuation occurred during the waxing and waning periods is certain: the only doubt is as regards the importance of these fluctuations. Some maintain that they were of minor importance, while others consider that periods of intermediate retreat occurred on so great a scale that these may be regarded as interglacial periods.

For general study of the glaciation it has been sufficient to regard the glacial period as one of culmination of the ice followed by retreat, but it must be remembered that the phenomena we have considered are equally explicable on the supposition that at least one interglacial period may have occurred between earlier and later periods of glaciation.

We may now consider such evidence as is presented by the glacial phenomena of the district bearing upon this subject. In doing so, it is important to take into account the possible importance of acts of erosion as well as those of accumulation in throwing light upon the question, as has been shewn by Prof. R. S. Tarr.

The occurrence of an interglacial period was advocated by Clifton Ward, Mackintosh and De Rance[1]. These writers believed that submergence occurred during the ice age, a supposition which is not considered by many at the present day to accord with the facts.

[1] Ward, J. C. and Mackintosh, D., *loc. cit.*; De Rance, C. E., "On the Two Glaciations of the Lake District," *Geological Magazine* (1871), vol. VIII, p. 107.

Mackintosh distinguished three sets of drifts on the plain of western Cumberland[1]:

(1) "Upper red loamy clay (partly derived from the waste of Permian strata)...contains few boulders."

(2) "Sand and gravel formation of various parts of the plain... contains pebbles and a few boulders of most, if not all, of the rocks found in the clay above and below."

(3) "Reddish or yellowish-brown lower boulder-clay, which is often varied by a light bluish or greenish tint, contains the greatest number of large boulders."

Bernard Smith accepts Mackintosh's subdivisions, and gives reasons for the following sequence of events:

(1) A period of maximum glaciation during which the Irish Sea ice and Lake District ice were confluent. The lower boulder-clay was formed at this time, chiefly by the Irish Sea ice on the coastal-plain, by the combined ice further inland, and by Lake District ice still further away from the plain.

(2) A stage of withdrawal of the ice "marked by the occurrence of moraines, moraine-terraces, trails of boulders and perched blocks."

(3) During this withdrawal the ice-dammed lakes and marginal channels of the west were cut, those south of the mouth of Eskdale following a slight re-advance of the ice "which overrode gravels previously accumulated as a marginal outwash-fan."

(4) The sands and gravels of the plain were deposited subsequently to the production of the western marginal channels. "The material was, in the main, deposited in embayments of the ice-margin, frequently beneath temporary sheets of water."

(5) "The upper boulder-clay seems to be the result of a restricted re-advance of the Irish Sea ice, which overrode the sands and gravels of the plain."

Apart from minor oscillations we see evidence here of a retreat of the Irish Sea ice after the period of maximum glaciation, and its subsequent advance during the formation of the upper boulder-clay.

Smith also gives reasons for concluding that the Lake District ice after the maximum glaciation withdrew concurrently with the Irish Sea ice.

There seems to be no direct evidence as to the amount of retreat of the ice during this period. It may have been a local phenomenon, or it may have been widespread.

When we enter the inner parts of the district the sequence is more obscure, owing to the general absence of the sands and gravels like those of the plain. Nevertheless, we may find some features which bear upon the question of a period of recession of ice between two periods of advance.

---

[1] *Geological Magazine* (1870), vol. VII, p. 567.

Though sands and gravels which are so abundant on the plain are rare in the heart of the district, we occasionally find such deposits between two sheets of boulder-clay. I have a record of a section near Elterwater Bridge, Langdale, about 8 feet high. It shews red boulder-clay at the base, and grey at the summit. At the right-hand side of the section the two come together, though there is a sharp line between them. At the other end, they are separated by a deposit of roughly stratified gravel. An isolated example tells us little, but similar occurrences should be looked for elsewhere in the interior of the district.

It has been shewn that the till which was produced during the period of maximum glaciation is best developed around the margins of the district, and that in the district it is most widely distributed on the low grounds of the Silurian rocks, but that in the higher grounds of the north and centre of the district it occurs in small sheltered patches in the mountains and may frequently be seen in the valleys. I suggested also that more of this material than is generally suspected may lie on the higher ridges.

If the valley-glaciers of the later stages of glaciation marked the final stage of the waning of the ice-sheet, one would expect that the drifts which during the period of maximum glaciation were being accumulated towards the periphery of the sheet, where the ice movement was feeblest, would during the later stages be laid down nearer the centre of the district. But if a re-advance of ice occurred, the drifts which were accumulated during the waning of the previous ice would be ploughed out by the ice on its re-advance, and relics only left here and there, as is actually the case. The moraine below Wolf Crag forms a line separating a drift-covered tract outside from a driftless one inside, as noticed by Ward, who regarded this as an indication of the formation of the moraine at sea-level during a period of submergence: the occurrence accords with the view that the glacier which formed this moraine ploughed out the pre-existing drift behind its moraine, and what seems to have taken place here may well have happened on a larger scale when the later glaciers descended the larger valleys.

Again, though as we have already seen moraines were formed during pauses in the recession of these later glaciers, they do not extend to the ends of the valleys. If the ice of the later glaciers was the dwindling relic of the ice-sheet, there seems no reason why these moraines should not extend far down the valleys beyond their actual downward limit. This suggests re-advance of the ice.

We may now regard the erosional effects to which reference was made. Tarr[1] shews that the hanging valleys of the Finger Lake Region were the result of two glaciations. In the first place, the main valleys were

---

[1] Tarr, R. S., "Watkins Glen and other Gorges of the Finger Lake Region of Central New York," *Popular Science Monthly*, May, 1906, p. 387.

deepened by ice-erosion, and then the ice retired.  During the period of retirement river gorges were cut at the mouths of the hanging valleys, *which do not extend to the bottom of the present valley.*  The ice re-advanced and its drifts filled these gorges, and this ice deepened afresh the main valley, so that the bottoms of the drift-filled gorges themselves "hang."

In the Lake District we find somewhat similar phenomena, but here the problem is more complex, and I mention these facts to direct the attention of observers to what may be a fruitful line of enquiry, rather than with a desire to argue that the evidence as it stands is in favour of two periods of glaciation.

On the east side of the Watendlath Valley several gorges seam the hill-side above the tarn.  That which enters the main valley close to the foot of the tarn has already been mentioned in connexion with altered rocks of the Borrowdale Series.  It is of further interest, inasmuch as it shews distinct proof that its upper part was buried in drift.  At one place a mass of drift still occupies the greater part of the valley, projecting nearly across it, so that the stream has to take a looped course to the north side of the gorge.

These gorges can hardly be pre-glacial.  It has been seen that in pre-glacial times, the district had an outline of subdued relief, and the occurrence of deep narrow steep-sided gorges is not in accordance with such relief, though they may be formed on a small scale as the result of local "cloud-bursts."

This particular gorge may however have been carved out by a stream flowing from the ice advancing from the Helvellyn range, which over-rode Armboth Fell, and during further advance the ice may have filled with drift the gorge which its marginal stream had previously eroded.

A large number of these drift-filled gorges are associated with hanging valleys.  In some cases the water in later times has taken its former course and partly or entirely cleared out the drift.  Sometimes the present course crosses the drift-filled valley, as below Small-Water, at a point considerably lower than that of the outlet.

In many examples the present stream-course and the drift-filled valley are parallel, though separated from one another.  Good examples are at the mouth of Gillercombe, and at that of the Seathwaite Tarn Valley above Duddon Vale.  In the case of these drift-filled valleys, as in those of the Finger Lakes, the floor of the drift-valley hangs as regards the main valley.  These hanging valleys were occupied by corrie-glaciers and it is possible that the phenomena were due to advance of these glaciers over the gorges of streams coming from their ends.

Taking all the evidence which has been hitherto collected, bearing upon the question of the occurrence of interglacial period or periods in Lakeland, one cannot regard it as giving any definite indication of anything beyond fluctuations of the ice on a fairly large scale, but it certainly does not appear to be antagonistic to the intervention of an interglacial period. The subject requires much further study.

Even if the large valley-glaciers were merely a stage in the waning of· the ice-sheet, it is possible that the latest corrie-glaciers were products of a distinct and very recent glaciation. Their relics are so fresh that we can hardly suppose a long period of time to have elapsed since their occupation of the combes. The existence of an actual glacier in a corrie on Ben Nevis was maintained a few years back: unfortunately it disappeared soon afterwards! But the occupation of a corrie by a mass of snow for many years in succession indicates that comparatively small meteorological changes might result in the formation of corrie glaciers in the uplands of Great Britain.

# CHAPTER XX

## POST-GLACIAL CHANGES

The effects of glaciation were to produce great changes in the topography of an area which had attained the condition of subdued relief, and one of the principal changes due to erosion was the steepening of valley sides and heads, which were thus converted into a state of unstable equilibrium as regards the ordinary erosive agents of the weather and running water. This unstable condition cannot remain indefinitely. After the ice had vanished, the normal erosion of a temperate region would be resumed, and its work would be in the direction of restoring the equilibrium which had been upset by glaciation. Accordingly one of the important changes that has occurred after the recession of the ice is a smoothening of the valley-sides. It no doubt began towards the end of glacial times as the ice receded, laying bare the tracts it had previously occupied, and has been continuing ever since.

Notwithstanding the general disturbance of equilibrium by glacial action, tracts were left here and there where the slopes were too slight to permit appreciable transport of weathered material, and this would remain upon the rocks from which it was derived. Gentle slopes of this character occur on many ridge-summits and also on valley-floors.

In the latter case, however, the flat tracts are frequently covered by material brought down from the steeper slopes above. On the ridge-summits this would not occur, and the waste would remain on the surface. It varies in coarseness according to the nature of the rock and the time during which weathering has gone on, and to some extent owing to the nature of the weathering agents. The most striking products of this action are the large blocks found on many hill-tops. Clifton Ward calls attention to "a certain *blocky* structure common to several of the mountain tops, and notably Scafell Pikes, which is of considerable interest, and gives great character to some of the scenic foregrounds." He regards the blocks as due to weathering but suggests the operation of earthquakes as a subsidiary cause. It seems unnecessary to invoke earthquake action, and the blocks are probably due to the action of frost assisted by other agents operating upon massive rocks with large well-formed joints. The name *Felsenmeer*, for which we have no English equivalent, has been given to a blocky surface of this nature.

R. A. Daly has described tracts shewing this feature on hill-tops of the North American mountains at the Forty-ninth Parallel[1]. He says of these tracts that "both as evidence of incomparably more rapid frost attack above the tree-line than below and as a condition for more effective attack by agents other than frost, the 'Felsenmeer' is significant."

It is unlikely that all the agents concerned in the smoothening process have operated with the same intensity since glacial times. There is reason to suppose that after the glacial period minor climatic oscillations have happened, which would give an impetus at one time to one agent, at another to a different one. Cold would intensify frost action, dryness would diminish the action of meteoric waters, humidity would increase the development of peat.

It is unnecessary here to discuss the various agents which cause the creep of waste material down the slopes: changes of temperature, water, not only as streams, but as inconstant runnels, the influence of animals, acting with gravity produce a downward movement of the waste. It will be sufficient if we notice the more important movements of waste which cause definite changes in the topographic features.

[1] Daly, R. A., *Memoir of the Geological Survey of Canada*, "Geology of the North American Cordillera at the Forty-ninth Parallel," Part II, p. 637 (1912).

The principal features produced by movement and re-accumulation of the waste are screes, landslips, dry deltas, snow-slope detritus, rain-wash and the alluvial deposits of rivers, lakes and estuaries.

Screes are the accumulations of loose angular detritus, which having been broken by weathering from the cliffs above, fal to the slopes beneath until they attain the angle of rest. They are universal in the district where cliffs are found. They generally form fringes for considerable distances below the lines of cliff, and when the supply of material is small, they are usually thin and do not necessarily extend far away from the base of the cliff: when owing to weakness of the rock or some other cause, more material is supplied, they are thicker and extend further away from the cliff-base. As a consequence the lower termination of the scree-accumulation generally has a scalloped outline. The upper surface is also usually irregular ; where the supply is large, the screes rise into the recesses of the cliffs in fan-shaped masses with the angles of the faces pointing upwards. The latter character is shewn in the accompanying illustration of the well-known Screes of Wastwater (Fig. 48).

Screes pass into dry deltas, where material is to a large extent brought down by inconstant water action of streams descending rakes and gullies.

Many of these screes are bare, while others are clothed with vegetation. It is usually inferred that the latter are older, and this may be correct. It is possible however that conditions in former times were more favourable for the growth of screes than those which prevail at present. This is a subject deserving further study.

Landslips are due to the slipping of rock masses of con-siderable bulk, and not as comparatively small detached fragments. The general cause of their production is well known. The rocks must be in a state of unstable equilibrium, and slides are prone to occur after excessive rainfall, when the material ready to slip is weighted by the water, which also lubricates the surface over which the slip takes place. As the glacial erosion produced a state of unstable equilibrium in many of the valley-sides, landslips have taken place with some

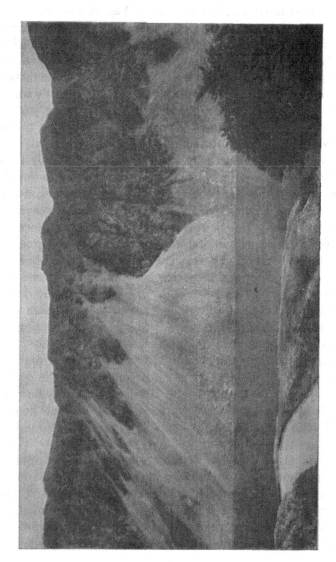

Fig. 48.   The Screes, Wastwater.

frequency in this district.   I saw a considerable landslip from a cliff composed of granophyre at the top of a combe north of Gale Fell, near Buttermere, in the Easter Vacation of 1900.   It had evidently taken place a few days previously, as the juniper bushes had been torn and bruised by the event.   Landslips generally leave a concave hollow in the place whence they come, and the material usually forms irregular mounds when it reaches its resting-place.   Landslips probably occurred frequently when combe-heads were exposed while their bases were being eroded by corrie-glaciers, and indeed landslip-action must be taken into account as a contributing factor to combe-formation.

Dry deltas are produced by small streams (often only running during heavy rain), which course down the hill-slopes. Some of these deltas are built at the bottoms of gorges, while others originate where springs issue from the waste-covered parts of hill-sides.   Deltas of the latter type are often formed by temporary streams bursting from the waste-covered ground, during those local heavy rain-falls known as cloud-bursts. William Gilpin describes the occurrence of a cloud-burst on Grasmer (Grassmoor) on Sept. 9th, 1760, which brought down a vast amount of material: this was spread out over the gentler slopes below.   The same writer describes a similar occurrence in the Vale of St John on Aug. 22nd, 1749[1].

Though lying somewhat beyond Lakeland proper, attention may be called to similar material accumulated in the gorge of the Lune below Tebay, on its left bank.   It was formed about the year 1858 in the course of three or four hours. "The débris still forms a striking object as seen from the train[2]."

The formation of these deltas, by streams piling up the débris on either side of their channel-centres, until the material is piled up to a height sufficient to cause the water to burst one of its banks, when it forms a similar channel in another direction, is well known.   In the young stage of a delta, these finger-shaped channels spread in a fan-like manner on the hill-slopes,

---

[1] Gilpin, W., *Observations on the Mountains and Lakes of Cumberland and Westmoreland*, vol. II, pp. 4, 36.

[2] Strahan, A., " The Geology of the Country around Kendal...," etc., p. 51.

separated by tracts of ground which have received no deltaic deposits. An incipient delta of this character is seen on the slopes of Lingmell, and forms a prominent feature from the Sty Head path. As the delta growth increases, and the stream course is shifted more and more, the intervals between these channels are ultimately filled in, and the normal fan-shaped delta is produced.

There is no difference between these dry deltas and those which are formed on lower ground, except that the latter have gentler slopes, are composed of finer material, and are often prevented from assuming the characteristic form, owing to the checking of their lateral growth by the steeper valley-sides. Much of the alluvium which occupies the valley-floors is of this deltaic character.

These deltas, as they increase in height, modify the topography by smoothing the hill outlines. They, like the screes, gradually grow upwards and mask the bases of cliffs. Furthermore they reduce the grades of steepened parts of stream courses, by accumulation at the base of cascades and waterfalls. A large delta is thus growing at the base of Taylor Gill Force in the Seathwaite Valley, and by covering up the base of the fall it has reduced its height. Similar features may be noticed in all parts of the district.

When a dry delta is formed near a watershed, it may produce a diversion of the stream which built it. A tributary stream flowing down a hill-side to a point near where two large valleys meet at a col will build up its delta in the valley into which it flows. As the delta increases in size, it may pond back the waters of the valley to form a lake, which will discharge over the col into the adjacent valley: or, without the formation of a lake, the delta itself may grow until it covers the watershed and sends a stream into the adjacent valley. A watershed-delta of this character is known as a corrom. If the grade of the head of the stream into which this diversion takes place be greater than that of the stream into which the delta-forming tributary originally flowed, erosion will lower the channel along the delta, and the diversion will be permanent.

A good example of a corrom occurs on Dunmail Raise[1]. A tributary, Raise Beck, descends with a steep grade from the Helvellyn range. It has built a delta across the watershed. The main drainage is now southward into the more steeply graded valley which leads to Grasmere, but in times of flood some of the water still flows northward. This indicates that the diversion is quite recent, and ultimately the entire drainage will be along the channel leading to the Grasmere Valley.

No doubt numerous other instances of corroms may be found.

The accumulations at the foot of a snow-slope may be regarded as special modifications of screes. They do not form now to any appreciable extent, but at the end of glacial times, when the glaciers had vanished, but the climate was still sufficiently rigorous to allow of extensive accumulations of snow during winter in the hollows, especially those of the combes, the fragments detached from the cliff above would skid down the snow-slopes to form a ridge of loose blocks at the base. As the general outline of these slopes would not differ in a marked degree from those of corrie-glaciers, the appearance of the ridge formed beneath the slopes would be similar to that of a corrie-glacier moraine, and the two would differ mainly in the characters of their component materials. I am not aware that any accumulations of this type have, in our district, been distinguished from moraines. It must always be taken into consideration that mounds which have an external resemblance to moraines are not necessarily the products of minor glaciers.

Rain-wash is a term applied to the finer waste spread over the valley sides and floors, due to the action of inconstant water-runnels carrying this material down the hill-sides. It is largely produced by the sifting of the finer material formed in the various ways described above, and by its downward transport. It usually possesses surfaces with concave curves on the lower parts of the valley-sides, and grades upwards into the screes and dry deltas above, and downwards into the

---

[1] See Oldham, R. D., "On the Origin of the Dunmail Raise...," *Quarterly Journal of the Geological Society* (1901), vol. LVII, p. 189.

river-alluvium of the valley-bottom, its surface becoming flatter as it approaches the latter.

Before considering the deposits which have accumulated on the valley-floors, we may pay attention to the formation of yet

Fig. 49. Goatswater, Coniston.

another series of tarns, which have partially or entirely come into being as the result of the ponding back of stream-waters by barriers formed of some of the post-glacial accumulations which we have been discussing.

We may notice two tarns which are blocked by screes. Goatswater below Coniston Old Man (Fig. 49) may be partly in a rock-basin, but its upper waters are held up by screes which have descended from Doe Crags. The screes at the lower end of the lake are sufficiently porous to allow the water to filter through them, and the tarn has no stream-exit except perhaps in times of flood. In ordinary seasons the water issues as a beck some yards below the tarn.

The little pool known as Hard Tarn is below Nethermost Pike, Helvellyn, in Ruthwaite Cove. It is a miniature tarn, being about 150 feet long and two to three feet in depth. Its interest lies in the fact that, though the normal exit is over screes, a wet-weather exit is over solid rock at the side. At one time the exit must have been over the screes alone: as the wet-weather exit deepens the bed it has already cut, it will become the normal exit, and for some time a wet-weather exit will be over the screes. Ultimately this will be deserted and the water will drain permanently over the solid rock. This tarn is fully described in my paper "Additional Notes on the Tarns of Lakeland."

Some tarns owe their origin either partly or entirely to barriers formed by dry deltas. Sty Head Tarn is probably in part a rock-basin, but it has increased in size and depth owing to a dry delta built across the valley by a stream coming down Aaron Slack, between Great and Green Gables. The stream now issues from the tarn between this delta and the solid rock on the other side of the valley.

A curious pool on the moorland south-west of Coniston village is found in the depression caused by the Stockdale Shales. It is known as Boo Tarn, and is now nearly filled in. It has been formed owing to two dry deltas built by streams entering the depression from the sides of Old Man. These deltas happen to be formed on either side of a low col, and accordingly the water has at times drained out at each end of the tarn.

Some of the tarns which seem to have been formed by a barrier of moraine may really be due to one of snow-slope detritus. As already stated, it is not always a simple matter to distinguish the one from the other.

I have met with no cases of landslip-blocked tarns in the district.

The alluvial deposits of the rivers call for little remark. It may be noted however that some of them may be fluvio-glacial alluvium deposited by rivers which issued from the ends of glaciers, and would belong therefore to the later part of the glacial period.

In places there is evidence that the rivers have lowered their beds since the earlier alluvium was deposited. Relics of this are found as terraces on the valley-sides.

Many of the river terraces of Lakeland are not carved out of rock below ancient alluvial deposits, but from drifts which had previously occupied the valley-floors. These terraces carved in aggraded matter differ from those of ordinary river-alluvium in some particulars, an account of which has been given by W. Morris Davis[1].

The alluvial deposits of the lakes need not be described in detail, but attention may be paid to the physiographical changes produced by them.

As the rivers enter the lakes they deposit their material and build up deltas. The largest deltas are usually at the heads of the lakes, and as the result of silting up of the lake-heads, several lakes have been considerably shortened. Long alluvial tracts are found above Derwentwater, Windermere, Ullswater and indeed all of the larger lakes, and the smaller lakelets and tarns, shew deltas at their upper ends. The actual delta of the lake is merely the end of this alluvial flat. Every important stream entering the lakes at their sides also builds up its delta, of which numerous examples can be seen in most of the larger and many of the smaller lakes. Ultimately an important stream may build its delta to such an extent as to push across the lake, thus separating it into two. This has almost happened in Haweswater, where the Measand Beck delta has pushed across to such an extent as to leave only a narrow strait between the upper and lower reaches of the lake. The process has been completed in the case of Buttermere and Crummock, which have now been severed by the delta built by a stream flowing from the east; accordingly we find the stream issuing from Buttermere flowing on the west side of the deltaic alluvium,

---

[1] Davis, W. M., "River Terraces in New England," *Bulletin of the Museum of Comparative Zoology*, Harvard, vol. v, p. 281.

between it and the solid rock of the west slopes of the valley (see Fig. 50).

Derwentwater has been separated from Bassenthwaite in the same way. Here, however, the delta is a double one, built by the Greta flowing from the east, and Newlands Beck from the south-west. Accordingly the stream which issues from Derwentwater does not flow between alluvial flat and rocky valley-side, but along the united edges of the two deltas.

Fig. 50. Buttermere and Crummock.
The river runs to the left of the delta separating the lakes.

Ultimately, if no further change occurred, all the lakes would be filled in, and converted into alluvial flats. This has happened in the case of some fairly large former lakes, and a host of smaller ones.

It is often inferred that a wide stretch of valley with a nearly flat alluvial floor marks the site of a former lake. This is not necessarily the case, for the flat may be entirely composed of river-alluvium. Many circumstances must be taken into account and sections of the deposits should be sought out and examined before finally concluding that such a flat occupies the site of a former lake.

The shingle-beaches along the lake shores may be briefly noticed. They usually have pebbles of less rounded outline than those of sea-beaches. Occasionally we find an island tied on to the mainland by a shingle spit. A good example occurs near the head of Crummock Water.

The beaches tend to form in embayments of the lake margins, and develop graceful curves in these embayments, as opposed to the angular outline of a lake formed by a barrier, in which shingle has not accumulated. An artificial example of the latter occurs at Tarn Hows near Coniston.

Beaches accumulate with unexpected rapidity and modify shore-lines in the way described. After the level of Thirlmere was raised, a few years only elapsed before beaches converted the somewhat unattractive angular indents of the margin into graceful curves[1].

The filling-in of the estuaries of the coast-line around Lakeland is generally similar in its effects to that of the lakes, save that it is modified by marine scour. Much of the material which is deposited in these estuaries is of the nature of alluvium brought down by the rivers of Lakeland, and the estuaries, like the lakes, were once filled with water to points considerably beyond their present upper limits.

The beaches of these estuaries are not unlike those of the lakes, for wave action is not pronounced, but tidal action produces some differences.

There is evidence that since some of the beaches were formed, a slight elevation of this coast has occurred, and we meet with raised beaches a few feet above present sea-level. A specimen of a raised beach from near Silverdale containing sea-shells is preserved in the Sedgwick Museum, Cambridge, and we have other evidence of their occurrence. A detailed study of them may elicit facts of interest.

We may now consider certain organic deposits of post-glacial date. The principal are the peat-accumulations, but shell-marls and diatomaceous deposits have also been found on the site of old lakes, and more should be looked for; they are probably not infrequent.

---

[1] See Oldham, R. D., "Beach Formation in the Thirlmere Reservoir" *Journal of the Manchester Geographical Society*, 1901.

Shell-marls are no doubt in process of formation in parts of the lakes at the present day, for many fresh-water molluscs inhabit them.    They have been formed in the lacustrine deposits of some of the filled-in lakes on the margins of the district.    The small sheet of water known as Haweswater, among the Carboniferous Limestone rocks near Silverdale, is the relic of a much larger sheet of water which has been converted into a peat moss.    Below the peat a shell-marl is found in places.    Similar marls may be expected to occur in some of the filled-in lake-basins in the interior of the district.    It is important that collections of shells in such deposits should be made, for some of the forms may be species no longer living in this country; furthermore they may possibly furnish indications of climatic changes.

Diatomaceous deposits are also being formed at the present day on the floors of some of the lakes, as for instance that of Windermere.    These again may be looked for among the deposits of filled-in lake-basins.    One such has been recorded by Dr Strahan in parts of the old lake which, until comparatively recently, stretched south of Kentmere Church, before it was artificially drained.    It is from this old mere that the valley is named.    The deposits occurred in patches below the peat. "Nearly 70 species of Diatomaceae have been determined from this deposit by Dr Stolterfoth of Chester[1]."

Peat occurs in those sites where the natural drainage is insufficient to carry off the surface-waters.    These conditions are found on many of the flatter ridges between the valleys; also on the sites of filled-in lakes and on the alluvial flats at the heads of the estuaries.    A subaqueous peat may be formed in shallow parts of lakes by aquatic plants.    The "Floating Island of Derwentwater" is apparently of this nature.    A work upon this island was written by G. J. Symons, F.R.S.[2]

Many of these peat deposits contain the remains of trees, even when at a high elevation.    There is evidence therefore that forests grew in places which are now devoid of them.    The

---

[1] "Geology of the Country around Kendal...," etc., p. 41.
[2] *The Floating Island of Derwentwater, its History and Mystery*, E. Stanford, 1889.

actual cause for the disappearance of these forests is still doubtful.

The peats of the ridges usually rest on glacial accumulations, and are therefore of later date. These accumulations of the ridges may however belong to a rather early stage of the glacial period, and the whole of the peats of the ridges need not therefore necessarily be post-glacial. It has however been shewn that many filled-in tarns were formed or partly formed by morainic barriers produced during the latest stage of glaciation, and the peats on the top of the deposits which have filled them in must therefore be post-glacial.

The study of the peats of various British districts by F. J. Lewis has produced interesting results with regard to variations in the floras of the peat at different horizons. Though no details have been given as to the succession of floras in the Lakeland peats, those of the Cross Fell district have been published[1].

The general sequence in that district is as follows:

1.  Recent peat: *Eriophorum, Sphagnum, Calluna.*
2.  Forest bed: *Betula alba, Alnus glutinosa, Lychnis diurna,* etc.
3.  Arctic bed: *Salix reticulata, S. arbuscula, Arctostaphylos alpina.*

The plants of the forest bed suggest temperate conditions replacing the arctic conditions of the lowest deposits.

It is probable that a similar sequence will be found among the deposits of Lakeland, and the condition there seems favourable to the elucidation of the problem as to whether the arctic peat beds of some tracts are contemporaneous with glacier-accumulations of others. This might be solved by a comparison of the peats of the ridges which escaped later glaciation with those of the corrie-tarns held up by the latest moraines. It is possible that deposits containing arctic plants may be found in the former and not in the latter. It must be remembered that many plants of alpine character still survive in the district, and the number of survivors was no doubt greater in going backwards towards glacial times.

There are various indications that in parts of the district at any rate the growth of peat has now ceased. Many of the peat mosses are shrinking, and becoming traversed by cracks, at the bases of which water collects, and saps the peat on either side, giving rise to wide channels separating the peat into more or

---

[1] "The Peat Moss Deposits in the Cross Fell, Caithness and Isle of Man Districts," *Report of the British Association* for 1907 (1908), p. 410.

less detached blocks. Over the sites of filled-in tarns these
channels become enlarged to a considerable degree by erosion.
The water at their bases is often lapped by waves which undercut the peat at the bottom, and cause its gradual destruction.
As the peat gradually disappears the filled-in tarns are restored.
In several cases the process is not quite completed, and mush-
room-shaped masses of peat are seen projecting above the
water, with undercut stalks. The process was first described

Fig. 51. Tarn on the Haystacks.

by Rastall and B. Smith in the case of a tarn on the Haystacks,
near Buttermere[1]; the figure (Fig. 51) shews this restored tarn.
All stages of the process may be seen in various parts of the
district, usually among pools on the higher plateaux.

The effects of erosion in post-glacial times demand some
notice. It has been seen that certain small gorges have been
carved by waters escaping from tarns blocked by the moraines

[1] Rastall, R. H. and Smith, Bernard, "Tarns on the Haystacks Mountain...," *Geological Magazine*, Decade v, vol. III, p. 411.

of the latest glaciation. Other gorges are unconnected with tarns. Where the gorges are most marked they have been carved along shatter-belts, and illustrate once again the importance of these as lines of weakness along which erosion has taken place with relative ease. It must be remembered that some of these gorges are pre-glacial, or at any rate antedate the final glaciation: they have been filled with drifts, which have subsequently been cleared out, as proved by the example behind Watendlath Tarn, where the clearance has not been completed. Many of them have probably not been filled with drift, but nevertheless may belong to the glacial period, having been formed by streams issuing from the vanishing ice during its recession. Of this nature are probably the gorges which are seen along the shatter-belts near Sty Head and Sprinkling Tarns, and many similar gorges elsewhere.

That erosion has taken place in post-glacial times is indicated by the formation of gorges above recent dry deltas, some of which have been formed by modern cloud-bursts, as for instance those near Tebay, and on Grasmoor. Of the latter, West remarks in his *Guide to the Lakes* that in one place the old channel was choked up, "and a new one cut open...through the middle of a large rock, four yards wide, and nine deep." In my paper on the "Waterways of English Lakeland" I called attention to a similar gorge which was cut through the Roman Road on High Street to the north of the summit. "The head of the ravine is a few yards above the road, and where it cuts the road it is about 18 feet deep and 103 feet across at the top. It is excavated partly in loose rubble, but largely through rock *in situ*, though much affected by weathering."

If changes of this character have been produced in historic times, no doubt much greater changes of a like character have occurred during the longer period which succeeded the disappearance of the ice.

The gorge of Scale Force is possibly post-glacial. Many of the waterfalls of the Lake District are due to glacial erosion, and the fall of the waters from hanging valleys into the main valley. No doubt Scale Force originated in this manner, but it is one of the few falls in the district which owes its present character to the contact of rocks of different degrees of hardness. It falls over the hard granophyre at the junction with the softer Skiddaw Slates; these slates have been readily eroded, and accordingly we find what for the Lake District is a comparatively rare occurrence; the vertical fall of the water down a considerable height.

The appearance of man in the district deserves brief notice. As no trace of palaeolithic man has been discovered here or anywhere in the neighbourhood, we are not confronted with the difficulty of ascertaining the relationship of the date of man's appearance and that of the glacial period. The only relics of prehistoric man belong to the neolithic, bronze and early iron ages. Various prehistoric implements have been found on the surface, dug from the peat mosses, or exhumed from burial places. Records of these will be found in Sir John Evans' works on *The Ancient Stone Implements of Great Britain* and *The Ancient Bronze Implements of Great Britain*.

Researches in the caverns of the Mountain Limestone tracts around the margin of the district will probably yield further evidence of man's occupation. The Kirkhead Cave on the Cartmel promontory gives evidence that it "was occupied by the Brit.-Welsh, and before them by the users of bronze, and possibly by a neolithic people[1]."

Further exploration of these caverns will no doubt furnish other remains of prehistoric man, and may throw some light on the contemporary faunas. J. W. Jackson has discovered an interesting fauna in a cavern known as Dog Holes on Warton Crag[2]. He found among the remains those of the Arctic and Norwegian lemming and of an arctic vole, also of a shell *Pyramidula ruderata*, only known in this country as a Pleistocene fossil. No human relics were found in this cave, but we may hope for their discovery when other caverns have been explored. They should also be looked for in the pot-holes of the mountain limestone, having been found in these in adjoining areas.

---

[1] Dawkins, W. Boyd, *Cave Hunting*, p. 125.
[2] Abstracts of the Proceedings of the Geological Society of London, 1910, p. 49.

When considering these human relics we pass definitely from the domain of the geologist to that of the archaeologist. There is however no break. Events have followed one another in our district as elsewhere from the time of the deposition of the early rocks to the present time. Many long periods are unrepresented by sedimentary deposits in the district, but by considering all the phenomena, not merely the deposits, but also the formation of the igneous rocks, and the consequences of erosion, we have been able to write a fairly connected history of the district from the time of deposition of the earliest Skiddaw Slates to the present day. There are it is true many gaps to be filled in. The elucidation of the geological history of a district is never completed. New discoveries suggest researches along other lines, and much still remains and will always remain to be done.

I have attempted to outline some of the problems which await solution, confident that whoever attempts that task will be rewarded, not alone by new discoveries, but by labours in a district which is generally recognised as one of the earth's fairest regions.

# INDEX

Printed in the United States
By Bookmasters